THE DEATH ZONE

BY

TERENCE J HENLEY

ISBN-13: 978-1505236293
ISBN-10: 1505236290

© **Terence J. Henley Nov 2014**
2 Eldred Close
Stoke Bishop
Bristol BS9 1DG UK
Email: terry.henley@hotmail.co.uk
Tel Home: 01179046506
Mob: +447863184026

THE DEATH ZONE

BY

TERENCE J HENLEY

DEDICATION

This book is dedicated to my wife, Angela,
my son Nicholas and my Daughter Gemma.
Thank you for helping me through the good
and bad times of illness.
Thank you for everything. All my love to you.

Other Titles by this Author

Miranda   Part 1 of the trilogy

The Miranda Gate   Part 2 of the trilogy

Miranda Revealed   Part 3 of the trilogy

Hazel's Pride

The Game

Visionary

A Note from the Author

To my avid readers, before you read this book, *'The Death Zone'* due to its content, I want to know this is a Science Fiction Story. The people in this story are part of a trilogy; the first Book *'The Game'*, is already out and doing very well, then I had lots of feedback asking me how this story started and the people in the universe got to be where they are and why women were put in charge of their ships.

This book starts with religion, it is the way the Utopians control the people in this universe the people on Earth. Religion is so prominent, that wars are fought over it and now the Utopians need an answer to a question, which is why they built this universe.

When you read this book, remember it is a Science Fiction story and set in another universe which as you will see is about an experiment. I just don't wish my followers to be upset and thinking I am putting any religion down, but on doing research about religion for this book, I have discovered that in the in the 9th century, the Arabs follow Islam, knew a great deal about science and technology, the Greeks were the same and going much further back, look at what the Egyptians did, how they built those huge pyramids and they still stand today. It is now thought among many scientists and religious scholars, that had the Arabs not been overpowered, we would have went to the moon in the 14th century and by now, we would be out among the stars.

So read the book and enjoy it as a story should be enjoyed. The final book in the trilogy will be out next year. I wish all readers a long and healthy life and most of all; enjoy the book.

PREFACE

On Utopia Prime, Godisious, one of the top scientists, has been given the job of trying to discover the way he can come up with an answer as to how they can best stop the Cycloves from attacking them and wiping out every species in their galaxy.

The Cycloves are a very angry and violent species that are seeking something, and if they cannot get the answer to their questions, they wipe out every species they encounter. They have been travelling from galaxy to galaxy, right through the known multiverse.

The Utopians are a very peaceful species but do have warships to protect them and other ships to explore space. They had already encountered one species that came to their galaxy and started to kill every species they came into contact with thinking the people were Cycloves. When the Utopians confronted them, they discovered who the true enemy was and know they will have to find a way to overthrow the Cycloves or at least stop them from wiping their species out.

The Utopians live for many thousands of years, and all have the ability of telepathy, teleportation and can use their minds to alter molecular structures. Being a peaceful society in general; they hope they can find a way to stop the Cycloves from attacking them.

Godisious, with his scientific team construct a huge Linear Hadron Collider a thousand kilometres in diameter and ten thousand kilometres long. Due to its size, they put it into orbit above Miran's largest moon, Hiabus, the fourth planet in their stellar system. Utopia Prime is the third planet and is also quite close to Miran, only an hour's flight using interplanetary speeds.

Halfway down the Hadron Collider is Port Troy; the entrance to the multiverse inside the Hadron Collider. It is a large triple airlock taking ships from the outside to the inside. To get inside it, the people and ships have to be miniaturised, a simple job for the Utopians as they have been miniaturising objects for tens of thousands of years. When the Hadron Collider is completed, they fire it up and send two God particles towards each other which collide and start the initial explosions to make a miniature universe. They do this a number a times making fifteen universes in total, one on top of the other inside the Hadron Collider.

They speed up the massive explosions and create thousands of galaxies inside the multiverse and Luciferious, Godisious's wife, helped the team create the planets and galaxies. Luciferious is a master at altering molecular structures and from the outside of the Hadron Collider; she concentrated on the matter inside and started to create miniature planets, stars and constellations. Within a month, the universes were complete and ready to have life forms added.

Luciferious came up with the idea of using a droid to put on the planets and programmed them to get them through their days and eventually get into space. By altering the texture of their bodies, they became sentient beings and made lives of their own with no soul, but they were given the idea that a God existed and they should believe in it.

Knowing what the Cycloves looked like through another species called Decidens, they created a duplicate of the Cycloves and put them in the thirteenth universe and locked it down so they could evolve and the Utopians could see what they were doing and how they acted.

In the third universe they placed the humans, the species that would resemble them and take their part in this experiment. Luciferious gave them biological bodies but all they did was move around and did nothing to progress and died very fast. What they needed was for the species to move out into space and do battle with the Cycloves, but it was not working. They had only seven years to discover the answer and they knew it would take time to get the humans into space. So Luciferious came up with an idea where they could possess the humans and move them in the direction they wanted.

Godisious set a large time differential from their world to Earth's so that one day on their planet was equal to three years inside the Hadron Collider. The earlier years on the planet were highly speeded up, where centuries on the inside were seconds in the real world.

With help from another species who were experts in constructing large buildings and vast complexes, they created a huge transfer station on the moon Hiabus. People went to the complex, lay on a couch and were connected to a state of the art health body system where the machine kept the body alive while the person was asleep.

The person would then send a part of them called the soul into a biological unit on the Earth. The soul would then be able control the unit and gradually get the units to create machines to fly, make weapons and then get them into space to eventually fight the Cycloves.

The only way in for the soul was through a tunnel and the soul joined the unit just as it was born and then left the unit and returned to the body when the unit died. The longest time a person would be in the unit would be two to three weeks, which would give the soul between sixty and a hundred years on Earth.

They also gave the units something to believe in that would give all the species in the universe a common base line, God. Luciferious knew that to get the biological units moving in the right direction, they would need hands on help on the planet and something to make them believe in God. So Luciferious and Godisious swapped names, Godisious being the main God, and now female would be on the outside of the Hadron Collider in a place they called Heaven, and Luciferious, the male, would be on Earth, or rather below the planet, living in a subterranean cavern which Luciferious constructed when she built the Earth.

There were also Zones they could go to and communicate with each other, but because they built the Hadron Collider very fast and needed to get it operational quickly, they did not think everything through and later would discover it was too late to add an interface that would allow them to send people in and out to Earth.

Luciferious, with two thousand scientists and engineers boarded the warship Viscount, commanded by Rear Admiral Marcus and entered the Hadron Collider to start a five year experiment to get the humans from Earth into a war-faring species that would travel into space and fight the Cycloves giving them a way to see if they could overpower them.

When the ship was small enough, it entered the airlock and made its way inside the Hadron Collider and moved down the side of the Hadron Collider to the third universe. It entered the universe and started its journey to the Sol system, where it shrunk again and finally landed on Earth at the South Pole where Luciferious entered the underground zone they called Haydes. During the following three Utopian years, more people were brought into the experiment by ship and some people went back home, but at all times there were always three ships from Utopia Prime inside the Hadron Collider within telepathic and laser communication distance to Luciferious so that if they needed an emergency extraction, a ship could collect them and take them out of the experiment.

Luciferious and his scientists were always giving the humans gifts and ideas on machines, but there came a time when no matter what they did, man was determined to go his own way and went against God. That was when Godisious sent her second son, Jesus, to Earth as a baby in the normal way everyone went to Earth, thinking it would help him turn the people back to God and move them forward.

After his crucifixion, when his soul returned to his own body on Hiabus, he returned to Earth on the warship Viscount and using an invisible lift system, Rear Admiral Marcus made it appear to Jesus' disciples that Jesus had ascended to Heaven. Although the people were now following God, it was not as they had hoped and when Luciferious gave the Arab scientist Ibn al-Haytham the ideas to get them off the ground and see the wonders of the universe, he thought they would very soon get into space. However, man intervened again and the Arabs lost their control over their technology.

Luciferious brought two great world wars to Earth in the 1900's and gave man weapons to battle with and man got to the Moon the mid 1960's. Luciferious was hoping they would advance very fast and gave a number of people dreams to help them design large spaceships but the current space agencies did not have the funding or believed in the people enough to build the ships.

It was then that Luciferious realised they should have made a way of getting them in and out of the Hadron Collider much easier. Mark, their eldest son had joined Luciferious after his father had been on Earth for two of his years, over a two thousand Earth years. Now it was the turn of Mary their daughter to come to Earth and she would have to get the humans into space or the entire Hadron Collider would be closed down and everyone inside killed.

As they did not really have a proper life, the Utopian Ethics Committee agreed that because this was an experiment, the species inside Hadron Collider would forfeit their lives. They would get all the people who had gone inside the Hadron Collider out by ship, and then quickly, over two days, wake all the people on Hiabus and the biological units on Earth would simply die.

The Hadron Collider could then be switched off and they would have to fight the Cycloves as best they could. It was now up to Mary to either get the humans into space, or bring them to an abrupt end. After talking to her mother and the Ethics Committee for many days, they decided the best way forward was for women to take control. All men wanted to do was fight battles on Earth for land, when they should have been in space, getting ready to fight the Cycloves.

Mary knew exactly what she had to do and how to get the humans into space, she would have to travel to Earth by ship; the same ship her father came in on. Like everyone else who entered Earth, except a few like her father and the scientists who travelled with him, she would forget where she had come from and only believe what had been put into her head. For her, Mary would think she was the daughter of God and Lucifer, her Father, also a God, but that he had done something bad and was now separated from his wife because he had taken control of the Earth.

It would be up to Mary to bring about a change to the people of Earth, put women in control and take them into space where they would in the first instance meet other species and join forces, giving them a large enough force to eventually do battle with the Cycloves who were now a force to be reckoned with. She would have only a hundred years to get the people of Earth ready and out into space. For this to happen, Luciferious would have to be shown up and put down, which would give him the chance to return home and help plan the battle that would very soon come.

CHAPTER 1

Outside Radstock church, hundreds of women, mainly female church ministers and Lay Preachers, waited patiently for the outcome of the crucial vote. Whether or not, women should be ordained into the Church of England and Anglican Church to give confession, mass and other services, until September 10th, 2019, only men had been allowed this privilege.

An earlier trial run in the early 1990's had been overruled by the new incoming church leaders. In their opinion, God had wanted man to rule His church; therefore, in His accordance, the earlier trials were finished. Women were no longer allowed into the inner sanctums of the church; neither would they be allowed any powers of the archbishop and other high offices.

In early 1995, women had lost the right to be a full church minister. Now they were fighting again for what they saw was their right, especially as signs had been given by the Lord that a second coming was nigh.

"I say women do not belong in the church as ministers at all; women should not preach and should never have been allowed to get as far as they have in the first place," shouted Bishop Saunders, from Bath.

"Women are a fine for dressing up the church and looking after the children in Sunday school, but that is as far as they should go. Bringing women into the pulpit can only do us harm; we will lose our dignity. The way we preach, with a harsh voice, because a harsh voice is called for to bring in the sinners, we get the word of the Lord through to these people. Women do not shout the name of God; they do not put over the word of God in a way which carries force; His force. We let people know He is our salvation, whereas women don't say it in the right way to get the word across," Canon Radford protested with a raised a fist.

"Please, let us not get carried away with emotion," Archbishop Langley replied trying to calm the situation.

"We have all seen the signs; they were predicted in the bible and the entire world knows the Lord is coming to us again very soon. I am sure this time He will not come as a babe in arms, but as a full grown man, ready to carry on His Father's will. From the signs, we know He will come at the end of this month; just two weeks away and the heavens will open. Open to what; women ministers? Are we going to allow a woman to be the first person to shake the hand of the Lord?" Canon Williams' voice echoed through the packed church.

He had hit a home truth; this vote was not to be just whether women should be allowed fully into the church; but whether a woman, should by chance be the first person in 2000 years to touch the hand of the Lord Jesus Christ. Silence filled the room; there was still the head of the church to consider their plight, King William V.

Although William had only been King for five years, he thought it was very clever how the church suggested the Queen abdicate in her time of ill health. The church had denied Charles to have the thrown due to his divorce and they did not wish to threaten the second coming in any way and upset God. This way, William could take the reins and lead the country into the second coming, a man at the head of the Church of England, not a woman. Of course, it was not the only reason church leaders wanted to do this: with a man now the head of the Church of England and the Pope in Rome, the women stood less of a chance to edge their way into the inner sanctums of the church.

It was man's destiny to meet and guard God's son; not women's. The signs were prominent that a man of the cloth would be the first to witness the coming of the son of God; man would not be pushed out of his own inheritance.

A mumble of voices turned to a loud, harsh cry of "No!"

"May I have silence please?" Archbishop Langley shouted above the ravenous storm of voices. "Silence! I pray for silence!" he repeated.

The cries of no diminished slowly as order was restored to the congregation. When silence again was within the walls, Archbishop Langley stood to face his ministers.

"We have a problem, a problem that refuses to go away; it continues to rear its ugly head to taunt us. We, the leaders of the Church of England and its associated churches must now consider very carefully, what has been brought to us here today. We must decide whether or not to allow women fully into our ranks. When we read the Old Testament, in Genesis it states God made man in his own image, and He called that man Adam. In the Garden of Eden, Adam was lonely; so God made Adam lie down to sleep and took a rib from Adam to make a woman, Eve. They roamed together in the Garden of Eden, until tempted by a serpent of the Devil; Eve took a bite from the apple of life and changed the way they were to be for all eternity.

"Adam and Eve had three sons, Cane, Able and Seth. The first man on Earth to walk free over our planet, we all agree was Adam and the first woman was Eve. It was through Eve that Adam was cast out of the Garden of Eden, made to work to live; break his back so that a woman could bear his children," he said and knew he had them on his side.

"This is our second chance to enter the Garden of Eden; to stop the flashing sword swinging back and forth baring our entrance. With the coming of the Lord, we will again be allowed into the Garden of Eden and man will walk free, no longer having to toil to feed his family; for the food of life shall be freely available to him. If we allow the intrusion of women into the church, and by accident or by fate, a woman should be the first to greet the son of God; then we will be forced out of the Garden of Eden forever; banished for all eternity, do we really wish this to happen? Do women wish this to happen? Don't we all wish to see the Garden of Eden again? It is possible this offer may never be made again; ever!"

The church was now completely silent; everyone was sat on the edge of their seat, listening intently to what Archbishop Langley was saying . . . the truth.

"That which is ours by right, shall never; never ever again, be shown to us. It is written in the bible there will be only one further coming of the Lord. The Lord will come to our planet only twice. The first time he was crucified on the cross to save all men; we must not crucify him again, especially before he arrives.

"With the coming of the Lord, there is also the Antichrist. We are most definitely my brothers, about to enter the Garden of Eden. A woman messed it up once before; we will not allow her to destroy our destiny a second time," he said strongly, hammering his fist into the pulpit.

A roar of approval erupted throughout the Church to a deafening pitch, and it was many minutes before Archbishop Langley could speak again.

"My brothers, not only will we meet the son of God, but the Antichrist will walk among us and we must not, I repeat, we must not be swayed by his false promises. I am sure that if we allow women into the church, not only will she meet the son of God, but she will also bring the Antichrist into our church as well; even if it is by accident, he will be here among us," Archbishop Langley shouted.

A roar of voices filled the church and 10 minutes passed before order could be restored. Archbishop Langley was smiling when Canon Matthews stood in the middle of the congregation.

"Archbishop, bishops, deans and my fellow brothers, you have known me for a long-standing member of the church, that I have given my time and effort to the Lord Jesus Christ. In fact, when the signs first came that the second coming was almost upon us, Archbishop Langley put me in charge of all research into the bible so that we, the members of the church would know for certain we would not be wrong."

Applause echoed throughout the church for all his work and devotion; but they also knew that many of the women preachers had called upon him to help in their research into the second coming.

"I was intrigued by Archbishop Langley's sermon." Silence filled the church and all eyes were now upon him. He paused, wiping his lips with his tongue, looking at the people around him gaining their attention.

"What the Archbishop has just said, is true. I'm afraid however; he did not go quite go far enough." If a pin had dropped he would have heard it as everyone held their breath, wondering what he was going to say next.

"The bible tells us Adam and Eve were the first two people on Earth, that after being thrown out of the Garden of Eden, they had three sons. That is where Archbishop Langley stopped, we all know through reading the bible what happened next, they went out and brought back a wife each. I put it to you, if Adam and Eve were the first two people on our planet and they had three sons; from where did they get their wives?" he asked boldly.

Whispers filled the church. Archbishop Langley was thinking of a way to silence him as fast as possible. These things should be said to him behind closed doors, not in the open before the people gathered inside his church.

"We must assume therefore, that our planet was populated while Adam and Eve were in the Garden of Eden and that the Garden of Eden lay outside the Earth. This means the Garden of Eden was somewhere else, either on another planet or in another plain or dimension. This is the only way the three sons could have chosen a wife each that were abundant in their time. I have read the entire Old Testament over a thousand times; I have explored the New Testament in great detail and I'm afraid I have to tell you, that there are a great many stories and fables that contradict each other. These are stories of a gigantic puzzle. Should we believe them all? Some of the fables could not have happened these are red herrings, which the bible is full of; just like the Garden of Eden. If we were to believe this fable, where did all the women and men come from; did they just appear? Where were Adam and Eve put on the Earth after being thrown out of the Garden of Eden and how long were they in the Garden of Eden: most of all, where is the Garden of Eden?

"I have been allowed; since the first sign was brought to us, to travel around the world and examine early scriptures and discuss matters with translators. I have managed to speak to modern translators who have looked again at the old scrolls; it appears some of the scriptures have been translated incorrectly and we have been misled," he shouted aloud, making his point heard.

A sound of people gasping sharply filled the congregation as they wondered what they were about hear. How, in 2000 years, could they have been misled about . . . God?

"I returned from Israel late last night and my trip was dogged with disasters. The plane was three hours late taking off, an engine caught fire as we crossed the Atlantic Ocean, we veered off course by strong winds and everything was against me. Here in England, thick fog on a clear autumn's evening stopped us landing at Heathrow and my plane was diverted on very low fuel to land in Edinburgh. Would you believe it, in our day and age of technology; leaves, leaves gentlemen, delayed my train by two hours! Then my taxi unexpectedly, ran out of petrol and to get here on time, I borrowed a bicycle, but on the way, it suffered not one, but four punctures. I took to my feet, and the heel of my right shoe broke off. No one was injured, but my journey to you this morning has been marred by problem after problem; it as if someone does not want me to get here to tell you what I have discovered."

"We are very sorry to hear of your troubles Canon Matthews, but we need to vote on whether or not we allow women in our church," Archbishop Langley interrupted.

"Do you see what I mean? It is being made harder and harder for me to say what I have discovered," Canon Matthew retorted.

"Then please continue, but time is of the essence please hurry," Archbishop Langley stated in an arrogant manner, much to the surprise of everyone assembled.

"God is among us here and now, in this very church, in this sanctuary we call His house. God is listening and I know it to be true; if we disappoint God, we will no doubt feel His wrath when His . . . rather, God's son comes among us."

"Do you know the exact date the son of God is coming?" a minister asked in a whisper.

"Yes I do; in a cave in Israel, I found parchments which tell of the coming of the Lord in the year 2027, on the 13th day of the eleventh month. His child will come to us at the sixth hour to walk among men and put right what has been wronged."

"The parchments, have you brought them with you?" a voice asked.

"Yes! I secretly got them out in this role, but they are old, very old and have to be handled with extreme care," Cannon Mathews stated.

People were now listening intently to what he had to say; even Archbishop Langley.

"I read the scrolls in English actually, but my guide, a Frenchman read them in French and Hebrew. I found another two men, a German, who was a very old man and could speak Latin as well and man from Holland and they all read the scrolls in their native tongue, we all agreed the scroll is from God."

He walked forward and placed his briefcase before Archbishop Langley. Opening it, he removed the scrolls very carefully and unrolled them. Archbishop Langley looked at him and the scroll and Canon Matthews addressed the congregation out aloud.

"My Lord Archbishop, you speak Latin, Greek, French and some German, tell this congregation of men; what language you see before you?"

Archbishop Langley glared at Canon Matthews then he looked at the scroll. "It is in English, it's a joke," he said and cleared his throat.

"Please, look again," Canon Matthews asked.

"Canon Matthews, you have taken this charade far enough. It is in . . . Latin. No, no, Greek, no, it is changing, it's in French."

"Actually Archbishop Langley, it's in Norwegian; it is my native tongue. It is quite plainly Norwegian, I can see it myself," Bishop Brown stated from the side.

"It is in every language, no matter what country you are from, you will read it as if it is in your own tongue. This is not a trick, this is the word of God, this is written in God's own hand," Cannon Matthews explained.

"It is a sign from God, a sign that we should not allow women into the church. This sign proves that we must vote against women. Keep them out of the church and out of the Garden of Eden so we may all reap its benefits," Archbishop Langley shouted at the top of voice, seeing what he was reading was indeed the word of God.

Shouts of, "Keep them out!" echoed throughout the church for the following ten minutes. During that time, Cannon Matthews argued with Archbishop Langley, until he was pulled aside by ushers. When the noise died down, Archbishop Langley spoke again to the congregation.

"I put it to you now that we vote on the matter in hand; we have seen proof from Cannon Matthews that the second coming will be in less than two days. Being so short a time, it will not be possible to ordain women ministers into the church anyway. I therefore put it to you, to vote against the amendment and stop forever, women joining the church.

"Further I say to you, that no woman is able to pass above the position of Sunday school teacher and all women currently higher than this position be immediately lowered to the position or asked to leave the church in a ministerial position altogether. Now, a show of hands for these new amendments," he shouted. Once again he hammered his fist firmly onto the table before him that held the scrolls, making his will felt to his audience.

At the back, hands shot into the air and then slowly, as ministers gave the matter some thought, they agreed with the motion. Ushers commenced the count, and after ten long minutes, those against got to raise their hand but only a handful of members disagreed and nobody abstained.

"I declare the ayes have it. Women will not be allowed into the church to preach," Archbishop Langley said with a smile on his face.

At that precise moment, a bolt of lightning struck the four lightning rods on top of the church tower. A clap of thunder, a might never heard before, rocked the church and the surrounding area sending thick clouds of dust, decades old from the high church rafters to fall on the ministers below. Windows rattled, ornaments shook, some rare objects worth thousands of pounds fell from tables and crashed to the floor, shattering into hundreds of minute pieces. The skies, a moment before bright blue, were now turning grey and black, with lightning flashing between the clouds over the women gathered outside Radstock church.

As heads turned skyward, Canon Matthews managed to break free from the two ushers, holding him. They were both shaking in their shoes as another clap of thunder, louder than before made the church rumble to its very foundations. Light fittings, rocked from side to side; throwing more dust over the ministers below. The sound of thunder echoed inside the church as people looked up to the roof, wondering what was happening.

"You don't understand what you have done," Cannon Matthews bellowed above the noise. The lightning, forming a circle high above the church, continued to flash violently between the clouds.

"You have made a great mistake, women belong in the church; they have more rights to the church than we have. It is written in the bible that they should succeed us. That's when men will . . ."

"Silence him!" Archbishop Langley commanded, pointing an outstretched hand at Canon Matthews. Then he held his hands aloft to God. "This is a sign from God that we have done the right thing. Someone bring the four women who represent the order into the church that I may speak to them," he bellowed.

Five minutes later, four women entered the church, looking about them as they walked down the central aisle. The vicars, bishops and ministers looked at the women as they passed them in silence. Once they were in the inner sanctum of the church, the black clouds overhead disappeared and the raging storm ended as abruptly as it had started. Archbishop Langley smiled as sunlight shone through the glass windows. Greeting the women with outstretched hands, he turned to face the altar.

Looking up to the rear of the church at the old stained glass windows, showing a scene of God and the Archangel Gabriel, Archbishop Langley's smile faded to a look of horror as he as he saw a crack from corner to corner crossing in the middle in a sign of an 'X'. The colour had trained from the glass so it was now clear; what was once a delight to look at; was nothing more than a ruin. The entire congregation followed his gaze; those who could not see the window, could hear the whispers moving back along the aisles of pews. The chatter grew to talk and talk to a wild frenzy of questions. Many of the young ministers, thinking they had done wrong, fell to their knees and prayed to God for repentance.

"Brothers, we have made our decision," Archbishop Langley said and smiled, lowering his face, turning to look at the four women standing before him. They could see quite clearly, tears had formed in his eyes and were now falling down his face.

"My sisters," he said wiping the tears from his cheeks. "We, the brethren of the church, have made our decision and we have decided in our wisdom, with the new knowledge at our hands, that the coming of the Lord is very close. It is with this thought in mind; we have decided not to allow women into the church organisation."

"No!" shouted Mrs. Peters, one of the four women leaders. Tears started to pour from her eyes and rolled down her face as she sobbed before Archbishop Langley.

"Pray silence please and allow me to finish," Archbishop replied raising his hands.

Mrs. Peters, with Mrs. King, Mrs. Meadow and Mrs. Rocc glared at the Archbishop. They were all lay preachers of the highest order, waiting to be ordained into the church. Their hopes were now flying out of the window and they were sure there was something wrong with the decision they had made. On seeing the desecration of the stained window before them, they knew no earthly hand could have done it. As they looked about the men in the front pews, they could see all of them were very shaken indeed. They knew Archbishop Langley was against women in general, and he did not want them in his church. He had made it clear on many occasions through papers and church magazines he had always been against women ministers.

"Now, I will explain why women are not to be allowed into the church. It is felt they do not have the character to push the word of the Lord forward into the ears of man; women do not push hard enough in their sermons to instruct men in the word of God. It was therefore decided by vote, women will no longer rise higher than Sunday school teachers and church helpers. Your positions as lay preachers will end from midnight tonight."

"Surely we are to be given time to talk with our colleagues and challenge your decision to take our review to the King of England, the head of the church?"

"From midnight tonight you are no longer welcome in the church as ministers or to teach the word of the Lord to people. You may teach the children in the ways of the church and the word of God, however, you will not teach the word of God to man; that is for us to do."

"Then the coming of the Lord is nigh, when is it exactly? We have a right to know the truth," Mrs. King demanded.

"You no longer have any rights within the church," replied Dean Headley, from the Archbishop's side.

"We are elected preachers until midnight tonight; you have just said so before everyone here. There are male lay preachers here within this congregation and they know the truth; we also wish to know the truth as we are also lay preachers, at this time. We only wish to know the truth, when is the Lord coming to us," Mrs. King continued to argue.

Archbishop Langley could see there was no getting away from telling them what they already knew; he had admitted they were still lay preachers until midnight. In his heart he knew they had a right to know the truth.

"Then you must keep this to yourselves; you will not disclose what I'm about to tell you to anyone outside of this church; do you agree to this?" Archbishop Langley asked the four women.

The church was suddenly silent, as everyone wanted to hear their reply.

"We do," they replied together after a short discussion.

"The coming of the Lord is in less than forty eight hours away. It will be at 6 a.m. on the 13th of November 2027."

The four women looked at Archbishop Langley, then at the other ministers sat by their sides. Their faces looked to the back of the church to the cracked stained-glass window. They knew only God could have done this to his own church, no other person would have dared desecrate this holy place. They knew that the men gathered before them had made the wrong decision, and very soon, God's son would tell them how wrong they were.

"Thank you Archbishop Langley for informing us of the time, we will not pass this information onto any of the women outside; however; we will stay close at hand. In our opinion you have made a grave error of judgment and only one person will be able to put it right."

"We have made the right decision," Archbishop Langley replied sternly. "Now leave this place and remain outside its doors until we welcome you into the church again to pray to God for your forgiveness."

The four women turned on their heels, and with tears filling their eyes, walked the length of the church in silence, their heads held high through the open doors that led to the congregation of women outside.

CHAPTER 2

When the church was finally cleared of clergy, a majority of the woman and their supporters had moved away from the main red doors as the four women were recalled to the church to speak with Archbishop Langley and Ian Huntley in the vestry to discuss the outcome of the vote.

In the commotion to clear the church, Canon Matthews had been released and was hiding in a corner unseen by the ushers as they got the last of the clergy on their way back home. When most of the ushers had left, Canon Matthews made his way to the vestry; he waited outside the door until the ushers were out of site, then knocked politely and boldly entered the small room closing the door behind him.

The six people gazed at him in silence; Mrs. Peters looked worried and Archbishop Langley frowned on seeing him. Dean Huntley stirred and went to step forward to escort Canon Matthews, from the vestry, but Canon Matthews, put his hand up to stop him and continued into the room, taking a seat at the table.

"Archbishop Langley, I know you didn't approve of my interruption, but you sent me to discover the truth; yet now you fear it."

"What is he talking about?" Mrs. Peters asked confused.

"Nothing! He doesn't know the truth himself," Dean Huntley answered.

"I have discovered the real truth about God, and neither of these gentlemen wish to hear it. It's now 6 p.m. in exactly thirty six hours; the second coming will take place. The Lord will again come to Earth; are you willing to take the chance that he would not come here first and tell you the truth himself? Then he may show you up before all of mankind, what would you do then?" Canon Matthews asked very annoyed.

"What do you mean; what is the truth? Do you know where the Lord will come first?" Mrs. Rocc asked urgently.

"I do not know where the Lord will come first, but I am sure he will come at this time," Canon Matthews replied, glancing at each person sat before him.

"At that moment, the vestry door opened and the bursar entered carrying a tray of coffee for the group. Upon seeing Canon Matthews, he stopped and was about to call for help.

"I am not here to cause trouble, just to tell the plain truth; put the tray down and please could you bring another cup?" Canon Matthews Instructed. The Dean nodded and a moment later, the bursar left, returning a minute later with another cup and saucer then closed the door behind him.

"The only way out of this situation is through it; so tell us what you have discovered, I can see you're not going to leave this matter alone until you have said your piece," Archbishop Langley said.

"Thank you. Most of the bible is true, but there are several parts which contradict each other. There are words which have in the past been translated not by exact meaning, but by fit, or appropriate meaning."

"What are you getting at?" Mrs. Rocc asked sitting forward on her seat.

"What we believe is that in the beginning, God created the heavens and Earth. We are slowly being turned in disbelief by scientists who say the universe was created by a catalytic Big Bang. We know that God is being pushed out of the church, out of our thoughts and our prayers." Dean Huntley argued.

"No Dean, I am talking much deeper. It says in the bible, God created man in his own image. We all have our image of God in our minds' eye. What do you see God as Mrs. Peters?" Canon Matthews asked.

"I suppose an old man with a long white beard, long white hair, a big kind man who lives and cares for his children. He would show love in his face and his smile. I imagine he is wearing a long white robe; because sometimes your mind mixes up God with the way we remember Jesus and the clothes he would have worn."

"It has long been asked if God is so caring; why does He allow war to persist, famine to continue year after year, making his children starve and die of malnutrition? Why can't he send rains to the desert regions, to Africa and allow his children to live instead of die in their mothers' arms? God could make the lands fertile instead of children being born into starvation. They have not sinned against God, except, perhaps by being born. Why does God allow diseases like cancer, AIDS and multitudes of other diseases we are plagued with, when all he would have to do is pass his hand over the Earth and bring good health to everyone? Surely by just doing this, everyone on Earth would turn to God and praise Him on high," Canon Matthews asked in a quiet manner. No one answered, but they all thought deeply on what he had said.

"We are all of the opinion that God in some form resembles us, or that we resemble him. Women believe in God more than men, yet here we are, throwing women out of the church. The decision today has broken the church wide apart. There will be many men who believe women play a vital part in the church and belong here.

"Sometimes we look for proverbs to explain what we want to get across to a congregation. Well I have a form of proverb, it's a joke in many ways, but it will suffice to show you how you can be wrong. To show you how you think.

"A boy was brought into hospital with serious injuries from a car crash. He was so poorly, a top brain surgeon was called for to operate and save the boy's life. As the boy was wheeled into the operating theatre, the top brain surgeon looked at the boy and said; I cannot operate on this boy, for he is my son. What relationship is the boy to the surgeon?" Canon Matthews asked and looked at the people sat at the table and raised his eyebrows in anticipation.

"He is the boy's father, naturally," the Dean replied smiling.

"Wrong!"

"Then the boy's stepfather?" Archbishop Langley added, sure that he was correct.

"Wrong again."

"Then the boy's adopted father?" Mrs. Peters suggested. She looked at Canon Matthews, trying to fathom the problem he had given them.

"Wrong yet again," Canon Matthews replied smiling. The room was silent; no one came up with another answer. "You cannot understand what the answer is and yet the answer is simple and you can't see it before your eyes. The top brain surgeon is the boy's mother."

"Ridiculous!" Dean Huntley protested.

"He's right, of course it's the boy's mother," Mrs. Meadow intervened.

"Have you forgotten that even in medicine, women get to the top? Yet we do not recognise them in our thoughts; so much so, that a simple question throws everyone into confusion. I say to you, beware of casting God in your own image." Canon Matthews looked around the table as he poured coffee into the cups and took his cup starting to drink the hot liquid.

"What are you saying?" Mrs. Peters asked urgently.

"What I am saying is that God did not make man in his own image. God made woman in Her own image." there was utter silence as the vestry door opened with two large ushers ready to throw Cannon Matthews out.

"Leave us immediately!" Archbishop Langley shouted, waving them out of the door. "Do you know what you are suggesting?" Archbishop Langley asked; leaning forward glaring at Canon Matthews after the ushers closed the door behind them.

Canon Matthews paused, looking at the church members. "Yes, we have made a grave error and took it for granted that God could only be a man and made us in His image. We read in the bible that He created Adam first, but who knows who He really created first, there was nobody there to record it and Adam or Eve would not have put it down in writing, I doubt they could even read or write. It is recorded in Genesis how they thought at the time, the beginning of man happened. Yet as I have said before, Cane, Able and Seth took themselves a wife each and it does not mention in the bible that the Earth was inhabited by other people. If we are to believe that Adam and Eve were the first two people on Earth and they had three sons, then life would have stopped when the sons died. There would have been no women to carry on their bloodline. So I ask you again, where did all the other people come from that inhabited our planet? We just accepted what the bible told us without question."

"But what of the Garden of Eden?" Archbishop Langley asked.

"It was more than likely there, but created by God for a testing ground. God probably put the Garden of Eden somewhere else, on another planet outside our solar system or in another dimension. He took Adam and Eve from Earth and placed them in the Garden of Eden to see how his subjects reacted; very much like we put animals in cages. It was an experiment to see what would happen, how man would be by himself. How woman would either help or destroy man. It is even doubtful that Adam was the first person created; most likely it was Eve." The women stared at Canon Matthews in disbelief.

"I don't understand," Mrs. King uttered. She was shaking visibly in her seat looking about the others sat around her, wondering if she was being told the truth or a complete pack of lies. As she thought on what Cannon Matthews had said, she started to see he may well be right.

"The Garden of Eden was an experiment to see how woman could survive without man. Then, when she desired a man, Adam was created. The tree was also placed in the garden to test man and woman; to see if man could keep a woman under his thumb. Eve was a woman with a mind of her own, like so many women of today; she desired the fruit and ate it, women were not cursed to bear children and men to toil in the fields because of the apple; it had been going on countless years before."

"But what of Jesus Christ; He cried out Father on the cross?"

"Another miss-translation probably; Jesus could well have called out mother. In the old scrolls which were translated, we believe God to be a man, therefore, it was a logical step to think he meant father as he was nailed to the cross and in a lot of pain. So the scroll was written with the word father instead of mother, it's a logical misunderstanding."

"Is there really a heaven then?" asked Mrs. Meadow, almost in a whisper.

"While I was in Israel I did some further research; there are a few men who say they can work miracles of healing; but when it comes down to it, women seem to be far better at healing than men.

"A man will lay his hands upon a person's head, but there is no heat; when women say they can heal, she has the spirit in her body. Her hands become hot, but not just warm; I mean hot enough to burn. The burning sensation travels through your skin and into your bones; it leaves no marks, but you feel, for a short time at least, better. The pain inside you will depart, as if it has been taken away. If you touch the skin where her hands were placed, you will find the area is hot.

"When we talk of ghosts roaming our planet for years, it is mainly troubled. But I have discovered many of these ghosts were firm believers in the church, it is thought that upon seeing the true God, they could not accept Her and would not enter the Kingdom of Heaven but walk upon the Earth until they believed in the real God."

"It would make sense, but who is the devil or is there a devil?" Archbishop Langley asked no one in particular.

"If you think God as a man and the devil as a man, who was together before a row broke out, we would have a man-to-man relationship; if God is a male person in the flesh, how could he have made woman? Why not make the male a self-reproductive model, surely God would want to be better than anyone else, to do this he would need to be able to make more of his own kind. Therefore, God would have made man in his own image, and there would be no need for women. How could God have imagined what a woman would look like if he had never seen a woman before? Only women have reproductive organs so it's logical to think that a woman created a woman.

"We make things better; improve on what we already have; if I said you have all the tools and materials at your disposal to make a Moolve, how long do you estimate it would take you to make one?" Canon Matthews asked.

"You have not explained what a Moolve is, so how can we make it," the Dean retorted.

"For your sake, let's say it's like this a cube of sugar," Cannon Mathews said holding the sugar cube in his hand. "This is called a Deelver; now you know what a Deelver is and looks like, what is a Moolve?" Cannon Matthews asked.

Silence filled the room; no one could answer the question, because no one understood what he was getting at.

"I feel this is a trick question; your Deelver is a cube sugar. We know that sugar comes from sugar cane. If I am reading you correctly, what you are trying to say is; what is a mate for the sugar cube?" Mrs. Peters suggested.

Cannon Matthew smiled and looked at her. "You are correct; to take it a step further, we would like to allow the sugar cube to reproduce, so what type of mate can we give the lump of sugar to mate with? It would be far easier to make the lump of sugar a single cell organism, an organism which will split itself into two and reproduce itself; similar to a simple earthworm. As you have been given the lump of sugar to be mated with, we could make a similar lump of sugar and call it a Moolve.

"If I were God, I would be a single cell organism; therefore, I would make the Deelver a single cell organism like myself. If God is male and He created man, then how would he know what a woman would like if there were not a female God to take his idea from? For us to be made in the image of God and we believe this to be right, He would have 'X' and 'Y' chromosomes in his body as we do."

"He could have been a single cell organism and matured from there," Archbishop Langley protested.

"Not possible; He would not be able to conceive a woman; He would have given Adam a friend to be with who was another male, a self reproductive male just like Him."

"If that were the case, we would all be homosexuals," Mrs. Rocc stated.

"If that were the case, there would be no women at all," Mrs. Meadow added.

"Putting it bluntly, there has always been a Mr. and Mrs. God!" Canon Matthews interrupted.

"Then what of Jesus?" Dean Huntley asked.

"We can make babies in test tubes; it's done outside the body. Mary was said to have been a virgin but she could still carry and deliver a baby. Mr. and Mrs. God made their baby, then like we use a test tube, they implanted the fertile egg inside Mary; very much like the surrogate mothers of our present time; the rest is history," Canon Matthews stated.

"Then who would Jesus be in the second coming?" Dean Huntley asked.

"Jesus could well be Josephine; we have always assumed we would be looking for a man, but I believe very soon we will see much more than the second coming. I believe that somehow, we will see God as Moses saw God," Canon Matthews explained.

"You asked a question much earlier about the suffering in this world. If God is female, why would She allow suffering?" Mrs. Meadow asked.

"I don't think she cares; I think She sees man making a mess of the world he tries to dominate. Man has tried to rule the planet while women wait in the background, a majority of the time anyway. The women worry, get hurt, but they have a far deeper understanding and belief in God than man ever had. It is this deep belief that has relieved them of pain and suffering.

"If you ask any woman what she wants out of a marriage, she will say love and understanding. Ask a man, it's a house, car, material objects, a decent job then children. A woman, however, could live with love alone; she could and does survive in a famine with very little food, as long as she has love. Yet a man would desire a plough and grain to feed his family. We have lost the necessities of life; we have built false images of God in our work and our progress with science. So much so, that we, or some people would make us believe, that the universe was created out of a Big Bang. Perhaps it was, but who created the positive and negative particles to allow a Big Bang to produce more and more sub atomic particles which make up our universe?

"The scientists have not yet answered this question; it is a question we should be asking of them. We should argue against the Big Bang theory, ask them where the particles originated from."

"God created the particles; that is what you are saying, isn't it Canon Matthews?" Mrs. King suggested.

"Of course I am and you have seen the light Mrs. King. That no matter how far back or forward in time we go with our science, they will never be able to explain where the particles originated, no matter what they say, God would be the ultimate answer. However, there were always two gods; one male the other female. It is this we have not conceived ourselves; it is this we have let slip us by."

"So where is the male God?" Mrs. Rocc asked, moving into the mainly one-sided conversation.

"Possibly the one man we fear the most, the man we call the devil."

"Then there is a hell?" Dean Huntley asked.

"I believe there is a hell, although not as we understand it to be, it could be a world of toil and slavery, a very hot place where we pay the price for our sins: unless of course, the parchment which I discovered tells a different story."

He picked up his briefcase and laid it on the table before him, everyone watched in anticipation as he unlocked it. They held their breath as Canon Matthews flipped open the catches with a loud thud, breaking the silence making everyone jump.

Mrs. Peters made a quick sign of the cross against her body and Mrs. Rocc did the same; Archbishop Langley found his hand creeping up his chest and gripped his gold cross firmly in his hand. Dean Huntley swallowed the lump in his throat and slipped his finger between his white starched collar and neck as Mrs. King and Mrs. Meadow gripped their crosses firmly in their hands as they felt a presence move among them. There seemed to be something else in the briefcase, something they all feared. Something they did not wish to see; the truth!

CHAPTER 3

When the briefcase was finally opened, everyone was aware that something else, a presence of some form, was now with them in the room. It was much colder and an icy chill was felt by everyone sending goose pimples up and down their spines. A cold breeze blew the heavy red, velvet curtains which were hiding a picture of Jesus on a cross, so that it was now visible to the people in the room.

The handle on the heavy oak door turned, but despite the weight of Hudson's body pressing against it, the door would not give way. A desperate, frantic knock sounded inside the room.

"Sir, are you all right?"Hudson's anxious voice asked.

"Yes Hudson, what happened out there?" the Archbishop called back.

"There is a furious wind blowing throughout the church; it has knocked over books and papers throwing them into the air. The ushers have all left in panic and I am the last man here; what do you want me to do?" he asked as he looked about the devastated church wondering what they had unleashed inside the vestry.

"You may leave the church, be not afraid, no harm will come to you I'm sure. We will all be fine inside this locked room, the Lord has work for us and I am sure He will not harm us if we are to do His will," Archbishop Langley replied in a calm collective voice.

"But Sir, I cannot open the door, have you locked it?"

"No, the Lord has, do not fear for us; leave us and we will be out when we have concluded our talks. I am sure we will know the answers we need very soon."

"Archbishop . . . Is the Lord here with us now?" Hudson asked.

At that moment a gust of wind, in the shape of a hand caught hold of Hudson and pushed him away from the door and out into the aisles where the hymn books and papers were still blowing about in the wind. He never heard the reply as he looked back towards the oak door, he was certain he was no longer allowed or wanted in the church.

As he tumbled towards the floor, he grabbed a pew and pulled himself to his knees, crawling along the floor; he slowly made his way to the giant oak doors at the rear of the church. With the wind getting fiercer he could no longer stand as it continued to evict him from the church. As he turned one last time on reaching the large oak doors, he saw a pew being lifted into the air and deliberately smashed against the floor until it was nothing more than firewood.

A crowd gathered outside the church doors and watched in wonder as Hudson, the church verger, crawled on his hands and knees through the inner oak doors. His black gown was blowing over his head in the ferocious wind that was evicting him from the church. As soon as his body was clear of the doors, they were closed with a loud bang and everyone heard them lock.

Getting his breathe back; he slowly stood and looked at the shocked faces staring at him. He managed to get to his feet, turned and tried to pull the oak doors open, but they refused to yield. Turning, he faced the crowd which consisted of several news crews and the women lay preachers. As he looked at their shocked faces, he wondered what they were looking at; they had not seen, he was sure, what had transpired inside the church.

People were now pointing at him and as he walked towards the main doors, he caught a glimpse of his image in the wall mirror which was put there for people, mainly women, to check their appearance before entering the church. He stopped, turned face on to the mirror and then stumbled forward, getting closer to the mirror, hardly believing what he could see. As he glared into the mirror only a metre from it, he could see his hair was brilliant white, when five minutes earlier, he had black hair. Not only was his hair white, but his eyebrows and stubble was also white; as he looked down at his arms, the black hairs on his arms had also turned brilliant white and he started to wonder if all his body hair was white.

The crowd parted as he turned and walked towards the main doors; as soon as he passed them, the deep red, metal doors closed with a loud bang and were locked behind him. Everyone heard the doors lock and they also knew there was no one else in the church that could have turned the key.

A deathly silence filled the crowd as they waited for Hudson to speak. He was looking around in disbelief, wondering why this was happening. He was just about to open his mouth when everyone turned their heads to a sound and looked up to the steeple as the single bell commenced to peel loudly in the belfry.

Questions were shouted at him by men and women asking what had happened inside the church and who was peeling the bell. Some women preachers asked if the Lord had entered the church and everyone wanted to know what had happened to turn his hair white in a matter of minutes.

He looked at the group in silence, his mind thinking of those locked in the vestry, wondering if they were still alive. Only they knew the real answers to what was happening here and inside the church, only they could unravel the truth which had transpired in the house of God.

"I believe something is inside the church, something not of this Earth. Whether it be the Lord or God Himself or both of them, I am sure, certain in fact, that one of them is within the church and need to speck to Archbishop Langley and his people," Hudson replied to the press and television crew who were filming him and holding a microphone close to his face. Flashes from cameras filled the area around him as press photographers got photos of Hudson.

"Did you see Him?" one man asked.

"What did He look like?" another press man shouted from behind.

"I saw no one, but the wind was furious and tore the church apart, something or someone was controlling the mighty wind and it evicted me from the church. Had I remained there, I would surely have died as I was not invited into the discussions," he replied.

The press were getting out their mobile phones and some were running to their cars and TV trucks. A news story was unravelling at tremendous speed and would need to go out on air as soon as possible.

***

"What is in your briefcase?" Archbishop Langley asked.

"Papers which show the way to Heaven is long and hard: that after death there is life, but not life as we know it or think it to be."

"So is there a Heaven?" Miss Rocc asked.

"Yes there is, but it takes time to get there and we may have to travel through many lives to achieve this. Every day of our lives we toil a little harder for the truth and to get there."

"What is the truth?" the Dean asked sitting forward in his chair.

"The truth, as far as I can understand, is that when we die, the part of us that we call the soul goes to another world, to a dimension which co-exists with our dimension. Heaven is in another dimension, only reached after death and our soul carries our thoughts, memories and emotions into the next plane."

"Do we keep our memories?" Mrs. Rocc asked looking concerned.

"It depends on the age of the soul, some people when they die, will not have finished what they were sent here to do, so their soul is reincarnated into another body and the soul lives again, hoping to do and achieve what it was meant to do the first time. Some of us have been here many lifetimes, some of us have only been here once, it depends on how much the person listens to what God is telling them they have to achieve in their lifetime.

When we go to Heaven, we do not age as fast as we do this side. Imagine a dog, its life is seven years for one of our years, for the soul, its year is equivalent to fifty of our years. So when we die, the soul can wait on the other side for a loved one, then they go on together, or in a family group if the love is very strong. Love is also another very important factor which allows the soul to remain in the between place for someone it has loved on Earth. Waiting for that person to die so they can move forward together and go on to the next level and travel to the next world hand in hand and still be deeply in love."

"I don't understand," Mrs. Meadow interrupted.

"When we die, others are born, I suppose the easiest way to put it to you, is that when a soul dies another soul is born into the life on our plane, in our dimension. It's given a chance to fulfil itself, to live again in a dimension which it detests, for this dimension, this Earth, this planet, this plane in which we live, is Hell itself."

"Hell?" Archbishop Langley questioned.

"It is where the soul can experience pleasure, greed, hate, fear, anxiety, its only hope is the Lord, and it's only way out, Death!" His words sounded cold and hard, but as he spoke, something inside them began to turn, to make them see he was telling them the awful truth.

"If I'm hearing you right, Hell is here on Earth, but surely Hell is supposed to be governed by the Devil, a man with a fork where it is hot all the time. People are supposed to cry in pain, wishing they had not been so stupid on Earth and punished others for their greed," Dean Huntley replied.

Cannon Mathews smiled. "Do men and women not cry here? Is it not hot, very hot at times and at others, very cold? In our equatorial deserts, the temperatures are extremely hot and we have been forced to move away from these uninhabitable places on Earth. Is it not terrifying when we come up against something we cannot defend ourselves against? We worry, we suffer, our bodies decay before our eyes, our bodies get hurt and suffer disease and malformation. Our organs get infected with diseases like cancer, something we cannot cure, yet. Our body's age and we lose our sight and hearing, our legs slow us down as we get older and our backs bend under the work we undertake through our years on this planet. With all our technology, we cannot stop the ageing process and we cannot stop us from dying.

"We cannot cure cancer, Aids, or many other diseases, if we lose a leg, arm, finger or toe, we cannot re-grow them. We cannot control our destiny, we have wars, we blow people to pieces, we kill others and for what, nothing, for we can't take it with us. If we all believed in God, no matter what name we call him, there would be no wars, but there would still be death, only God Herself can change that. The God of Love," Cannon Mathews replied.

"You say God of Love, but surely this God would be only too pleased to help us here on Earth?" Archbishop Langley questioned.

"Would She? As I said before, man only needs material objects; he has them in his hands right here on Earth. Women can survive on love alone, if you ask any women if she wants war, she will say no."

"So will men," Archbishop Langley interrupted.

"Most men," Cannon Mathews corrected him. "There are men in the army, navy, air force who are there to fight and many enjoy fighting. They wish to fight to show they can be a hero, and then there are those who wish to control countries and the people who live in it and take great pride in watching their people suffer under them. People in government play at war, yet the two main powers, America and Russia are now at peace; Russia was demolished and split up by our way of thinking. Yet we didn't help them with food, factories to make clothes, televisions, cars, things we have in abundance; we managed to destroy Russia, but what has it left?" Cannon Mathews asked.

"Peace!" Mrs Meadow replied.

"And poverty," Cannon Mathews added. "This country still spends millions a year on armies for our defence. Why don't they stop spending money on ships and planes and build schools and hospitals, homes for people in this country? Why not help other countries around the world? No, this country still wants to be ready for war and to defend itself yet its leaders say they only want peace. Can you see how the Devil is among us? He will induce fear and hope but does not induce charity.

"Charity comes from love, affection and gratitude, just look at the companies in this country which make millions in profit each year. It s through greed alone they continue to put up the price of food and their products, so they make more money for their backers and put people deeper in debt." Canon Matthews said.

"The church is rich, very rich," Mrs. Rocc added.

"Yes it is, instead of building churches which cost thousands why not build a school or a hospital where the sick and needy can be helped to get well or educated?" Cannon Mathews asked.

"What would you do, have people pray in a hospital every Sunday?" Archbishop Langley asked laughing. "We need our churches to bring people to them so they hear the word of God through our ministers. We can't do this in a hospital or school," he continued.

"Why not? Look at a business, any business; a place of work where money is earned by making items or selling them. Within the factories and offices, people meet in different rooms; they make deals, exchange money and talk; the manager's talk to their workers. We talk to our flock, preach to them, we don't need a building the size of church to speak to people, preach, it can be done in any restaurant or workplace, or a small church in a hospital, we even pray now in people's front rooms during the week. We can preach in a school hall, where it has been done for centuries; so tell me, what is the difference? Why not help build a school or hospital and use one of the rooms for people to come to us where they can pray, listen to us preach; we could even use school halls on a Sunday to hold our prayer sessions there, so the schools will be used in two ways?" Cannon Mathews asked looking at each person in turn, leaving Archbishop Langley until last, his eyes resting on his, making him think what they had been doing, and what Jesus had to preach from.

They remained silent but thought deeply; Archbishop Langley could see what he was getting at, what he was making them think. The pomp and circumstance which surrounded the church was indeed a waste of money; the church was only really in use one or two days a week; some were open a few lunchtimes or evenings, but they were not used to their full extent. They didn't house the homeless at night and didn't treat the sick or teach the children how to read and write.

The money collected to rebuild churches was not really needed; it could have been spent on hospitals, homes for the elderly, schools or for the poor. Despite being good, he could see the church wanted material things and gains as well as people. Land, possessions, gold, great organs when before a simple small organ or piano would suffice. Did playing an organ very loud make the people sing any louder? It didn't, the sound of the organ drowned out the small number of voices and the people were diminishing by the week; it was getting harder and harder to get new blood into the church and fill the empty pews.

Holding his head low, he clutched his gold cross, suddenly realising how much wealth it contained. The red rubies, blue sapphires, green emeralds and diamonds reflected the light in the room, the cross and thirty inch heavy gold chain was worth £39,000.00. The safe it would return to when the meeting was over held more gold crosses, diamond rings, rubies, emeralds worth thousands and gold by kilos.

The articles in the safe were insured for over a million pounds and that was one safe, his safe. The main church safe, the large safe in the basement held the main bulk of gold and diamonds, worth in this church alone over three million pounds and this was reflected in every church in the country. The land and buildings sent their wealth into the billions, the church was indeed rich, and each week it begged for money to replace a roof or help the sick in Africa, yet they could sell the articles in one safe and build a new hospital or feed a tribe in Africa for an entire year. The church would never sell its gold, it didn't want to do all the work, it wanted the people to do it for them, was this what God wanted from His churches?

He thought of the size of the church, its spire and bell, the size of the church hall, the money it cost to heat in winter. For a Sunday alone, with the heating at full blast on an icy cold day, would be equivalent to heating a three bedroom house for two weeks. They wasted money on drawing people to this huge empty cold church where people would come and go. Doors would be left open allowing the heat out into the cold air; the rest of the heat escaped through the roof and single glazed badly fitted windows. It was wasteful; this he could see was what Cannon Mathews was talking about.

"If you're right; which I'm not saying at the moment you are, Mr. God is what we would call the Devil, is that what you're saying?" Archbishop Langley asked.

"Yes it is; we assume there is a devil because we think there must be an opposite of a positive, a good to a bad. In the bible it mentions the devil, but what happened before Jesus was born, did everyone go to Hell? Was there no Heaven until Jesus came into our world? It says Jesus came into the world, but from where did he come from? To come into something, then he had to come out of something else, didn't he?

"It is as it always was, when people died, they went to Heaven, but we had lost our way and God decided we needed something to make us believe in Him and pray to Him, to make Him great again as He was in ancient times. We sometimes say Hell is here on Earth, we were right, we do live in Hell. This world is a hell of place to live at anytime throughout our history. Tyrants have tried to rule the world, great wars have been fought and throughout history, there has always been poverty. It's the one thing which has been a constant throughout all time, poverty is the lowest of low and there are thousands, millions of people there all the time and we do nothing for them."

"I can see that," said Dean Huntley. "I was there myself, I know what it's like to be born in a cold house, have no money for food, no shoes on my feet, clothes on my back and the meals we would miss because we didn't have enough money to buy food. When I was old enough, I had to go to the woods to collect wood to heat the house and that was before I went to school. I promised myself then, I would change things when I grew up, but I never did, I just added to them," he admitted.

"We all have the power to change things, it is within us all, but for others it's a harder battle. When you have the things you once lacked, you forget the times of hunger and cold; you don't want to upset the apple cart; you don't want to tumble back to the depths from which you crawled out on your hands and knees," Cannon Mathews replied.

Everyone was silent; the women had fought for so long against all odds to get what they wanted, had in their quest stamped other people into the ground. They had used money destined to purchase food for the poor and needy, to prove their point and push them higher up the ladder. Even they had succumbed to the temptation of the Devil.

"Have you any idea what will happen on the day the Lord comes unto us?" Archbishop Langley asked.

"We will come to that when we have sorted out the truth," Cannon Mathews replied.

Everyone looked at him astonished, they could see he was telling the truth, he was calm and in complete control while they were falling to pieces. Things they had believed in for years, even preached to others were all lies; if they couldn't handle the truth, how could they expect the man or woman in the street to react?

Cannon Mathews paused for a long moment, and then refilled their coffee cups. As he passed each cup around, the cup was taken without question. When Mrs. Meadow sipped her coffee, she found it sweet, but not a sweet, sweet; it was she thought to herself, like nectar of the Gods. She took another sip of the coffee and realised the coffee jug had been emptied an hour before.

This was hot and fresh, yet nobody had refilled the jug; she took another sip then another, the jug of coffee was within her reach. Cannon Mathews had poured seven cups and she knew it would be almost empty. Reaching out she grabbed the jug with one hand and placed her cup on its saucer with the other. In one swift movement, the jug was before her and as she looked inside, she saw it was full of hot, sweet coffee.

"Would you care for another cup?" Cannon Mathews asked her.

"It . . . it's full," she gasped.

"We are in the house of God are we not? This is a special meeting, She will not allow us to go thirsty," he replied.

The jug was passed again from person to person and the coffee tried again by everyone. No one could find any fault with the substance before them; neither could they explain how it had come to be there, they had to accept the truth; it was a small miracle of God, but it was indeed a miracle.

"Now we are refreshed, I'll continue," Cannon Mathews stated after refilling his own cup. When Mrs. Meadow looked, the jug was filled again. "You may help yourselves to coffee as often as you wish," Cannon Mathews said with pride. He sat back and read a parchment before continuing.

"In the past we have always treated women as second best, they have remained at home while the men went to work, but at home, women slaved, cleaned and worked many more hours than the men ever did. They cooked, washed our clothes and bedding, taught our babies to become children and then taught them to read and write and taught them right from wrong. They washed dishes, put them away and kept the house and garden in some cases, clean and tidy, and their day started before the men got up and went to work and ended a long time after they arrived back home and went to bed.

"As their children grow up, they listen to their stories about school, help with their homework and clean their cuts and grazes, becoming a nurse when their children and husband become ill. All this happens while the husband usually sleeps through the night, of course he has a full day's work, but the woman is at home, working hour after hour, day and night.

"As more and more men have lost their jobs due to the recession and helped in the house, the women have got jobs instead and the men have discovered just how hard a woman's life is. However, we have always and still do put women down; now is the time for all women to stand up and be counted. They will stand before us no longer being beneath us; but our superior. Women have to be ordained into the church, even to the high position of archbishop." Cannon Mathews stared at Archbishop Langley who was looking devastated.

"Why not, don't you think a woman is capable of doing your job? Does her brain not equal that of yours? Does her courage, strength and feelings not equal that of yours, after all; it's not as if you do a manual job lifting heavy sacks of coal on your back all day long?" Cannon Mathews asked firmly.

"But . . . archbishop?" he stammered.

"For over ten years this country had a woman Prime Minister, the men of this country were given three chances to kick her out and when she left, the men did it underhanded while she was in France and her own cabinet stabbed her in the back, they should be ashamed of themselves. It was not her idea only of the tax that brought her down, and she had men in her cabinet, it was these people who brought her down and the country kicked them out as soon as they could.

"She stood by her convictions even when they were wrong, she admitted it and tried to do something about it to rectify it. The leaders now are a lost cause, like the church we too are lost and we need new blood to lead us and bring about the second coming. We must allow women to stand by our side and be our equals, only then will we see the Lord and the second coming will come to be."

"Are you saying that if we deny women to stand by our side and be our equals within the church, the Lord will not come and the second coming will cease to be?" Archbishop Langley asked bewildered.

"She may not come to us but appear to the women of this planet and take them away from us. What will we do then; man will die, we could not reproduce and even if it took a hundred years, we would all die, alone."

"Surely it would be the same for the women?" Mrs. King asked.

"I'm sure if the Lady took them away, she would ensure they would survive, don't you?" he replied smiling at her.

"More importantly, man's soul would go to Heaven, you said so yourself," Dean Huntley said urgently.

"I said nothing of the sort; I said the soul goes to another plain, another dimension. I did not say it was Heaven, I said the road to Heaven is long and hard."

"Then please explain this more clearly," Archbishop Langley demanded.

"When we die, our soul travels to another dimension, the in between."

"In between what?" questioned Mrs. Meadow.

"It is the place that lies between Heaven and Earth; there are other dimensions, places that people can go to and live in, places that overlap other dimensions so that on Earth, there are other Earth's, maybe two or three, but separated by dimensions, thin layers of space that separate the lives of the people who live there, we can't see them and they can't see us. Heaven is in another dimension and may occupy the same place that we occupy, but we can't get to it unless we die.

"When we die, our soul traverses the space between dimensions and can get trapped there, or decide to remain there and wait for a long life partner before travelling on to the next dimension. When we die, God pulls our soul from our body and through the light to Heaven, if we have fulfilled our destiny here on Earth. Or we may have to return in another body to do more before being allowed to move on and eventually stay in the house of God.

"It was once said that Heaven must be a big place for everyone to be in, but that is not quite right. There are souls, more souls granted each year and sometimes, a soul may find itself in the body of another person, a man previously may then occupy the body of a women fifty or a hundred years later to complete its journey before it has learned enough to move onto Heaven.

"Think of a soul as human being, when it's born, as a baby, it has to learn to stand up, eat, read write and grow to live with other people. A soul has to do the same thing only as I said before, the life cycle of soul is not the same as a human, its life is slower, but it has to learn how to grow and live with other souls, it is through being on Earth, living with other people that the soul learns this experience."

"The resurrection, being born again, images of a past life, multiple lives," mumbled Mrs. Rocc.

"What are you jabbering about?" Dean Huntley asked sounding annoyed.

"There are people who say they have lived here before, they are ridiculed by us and even when they are put under deep hypnosis, we say and tell everyone it's all a pack of lies. We don't want to believe them because we are frightened they may be right and what if we have been living multiple lives, what does it mean for us as members of the church?"

"Mrs. Rocc; if you're not going to explain your feelings, please do not speak," Dean Huntley shouted.

"I know about these people," Archbishop Langley interrupted. "A person called David Fletcher came to me but a week ago. He had been troubled by strange dreams of a past existence; in fact, he thinks he has been here no less than fourteen times. Most of the time he has been here for short periods, not over twenty five years, but it was always in war and poverty. His most adventurous time was when he was a Roundhead; he told me he killed over thirty men in the Wars of the Roses and he can remember every detail as if it happened only yesterday. He can even recall how the gunpowder smelt and the rotting flesh of men who had been left to die in dirt and mud."

"Obviously he has read books on the subject," Dean Huntley butted in.

He is diagnosed as dyslexic; he cannot read properly and never took history as a subject in school."

"Stories then, he watches the history programme on television, it could be anything like that."

"I would have said that as well except for one thing, and this one thing makes me believe him. David Fletcher is five years old; he is blind in one eye and has a scar. He was born with the scar which according to his hospital doctors and researchers, the scar could only have be caused trough a pike being pushed right through his body."

There was a sudden gasp then silence. Archbishop Langley looked around at everyone.

"The smallest scar is on his stomach, the larger scar is on his back, where the pike would have emerged and took a lot of his back and spine with it as the pike travelled through his body. MRI scans show positive scar tissue and damage to his spine, although his spine is fine and working, the MRI shows that at some time in the past, his spine was broken into six pieces as the pike travelled through his body. There is also scar tissue on his organs as well.

"He is blind because he has no left eye at all; it's as if his eye was plucked from its socket. I have read the transcripts of what he has told his doctors under hypnosis and what he knows himself. He has told his mother, father and doctors, he recalls lying in a field with the battle almost over.

"There were large crows and other birds of prey circling overhead looking down upon the dead and dying soldiers in the battlefield. He was bleeding to death and couldn't move his lower body and left arm, he was suffering from severe shock when a bird came down and landed on his head. For five long minutes it was pecking at his left eye until it eventually plucked it out. He was writhing in agony and lived through the entire gruesome event. Another soldier saw it happen and threw a stone at the bird, the stone hit David and the bird flew away with his eye. The mark where the stone hit his face is still there today, he also has signs of a previous broken cheek bone, which is where the stone hit his face."

"Do you believe his story?" Cannon Mathews asked.

"When I spoke to his doctor and saw the scans and x-rays, I had no option, something happened to him in a previous life. I was finding it hard to believe his story until what you have said; now I am beginning to wonder how many lives we have to live on Earth to get to Heaven," Archbishop Langley replied.

"There is an answer, but I'm not at liberty to tell you at the moment," Cannon Mathews replied.

"To return to what I was trying to explain, in principle, you die, move along to the other side and then you are born again. It is a mode of the soul to keep moving, usually with a lifelong partner who they search for on Earth and usually meet. Each time the soul gets more experience, but God in Her wisdom, does have mercy for some. People here are put through Hell for a specific purpose," Cannon Mathews continued.

"What are we in, some kind of zoo?" Dean Huntley asked.

"I suppose you could put it something like that, if that is how you perceive life."

"What of these ghosts, poltergeists who frighten people, where do they come from or reside?" asked Mrs. King.

"These are souls that die either in horrific or tragic circumstances, where things should have been said. A sudden death, a death which God or the Devil had not foreseen to happen and things had been left unsaid, things left undone. In some cases, although our paths are laid down for us, we have roads to choose, choices to make. Some of these roads are for good, others bad, sometimes, even God makes a mistake that can't be put right.

"That is where the ghosts come in, they take fate in their own hands, they may commit suicide, run accidentally in front of a lorry, a bus or train to end their life and it was not the way God had intended. These ghosts don't find themselves in the usual dimension of souls, but another dimension which co-exists with Earth. It's possible to transcend between the two, which some people do?"

"It would be horrific," Mrs. Rocc admitted. "You're referring to mediums and priests who try and pray for the soul to go over into the next life if you have the psychic energy to move between the two planes and these ghosts absorb the energy in their dimension, this allows them to move between dimensions and interact in ours, show themselves, speak to us, destroy items, touch and make us feel icy cold. They are locked in a dimension God has no access to or has decided to leave it alone for their penance of not abiding by His or Her laws. If you were locked away for all eternity, wouldn't you go insane and try to make contact with people on the other side; to try and get them to help you move on?" she continued.

"Yes I would, the souls are trapped there and the only escape is through the church. Neither

God nor the Devil wanted the souls to go there, but this is where some end up. The problem is, we, the church have made it appear God is in the image of man and when they see the truth they can't take it, so that is where they stay. They are locked in their time line and space, in the distant past, but follow us like the moving hands on a clock."

"Time! Trapped in time, it sounds like science fiction," said Dean Huntley sniggering.

"But nevertheless true," Cannon Mathews retorted.

Silence again filled the room and Cannon Mathews refilled their cups with fresh, hot coffee; from a covered tray which no one could recall seeing before, he passed around fresh sandwiches and cakes as sweet as the coffee. Half an hour later the group sat back and listened again to Cannon Mathews; this time they believed what he said; now they too wished they had accompanied him to see what he had seen with his own eyes.

CHAPTER 4

On the streets, the story surrounding the happenings at Radstock Church and Mr. Hudson seeing God was now on the news. It was also in the papers which were selling fast and furious.

All the roads surrounding the church were blocked and police had tried to break in through the main and side doors of the church several times to no avail. No matter what force they applied, nothing happened. The strange wind continued to encircle the spire and the bell continued to toll.

Mr. Hudson had been questioned by the press and the police who had released him as they could not get into the church to see what had transpired. Mr. Hudson was now in a car being rushed to a television studio where he was about to go on air. He was shaking from head to foot and his hair was still whiter than white.

In the studio, several background curtains were changed to show the colour of his hair better in the light. Other members of the television crew had raided his house, searching for a recent photo of Mr. Hudson and the people locked in the church. John, the producer, was counting the seconds off and waved his hand at a man sat in an easy chair.

"We have broken into our normal programmes to bring you this up to date information on the strange happenings at Radstock Church. My name is Mike Adams and with me is Mr. Hudson, the current verger at Radstock Church. Earlier today, a vote was taken by members of the high church and led by Archbishop Langley, not allowing women into the church to preach the word of God. It was also agreed by Archbishop Langley that women would no longer be allowed to preach as lay preachers. Their only teaching sanction is now through Sunday school, to look after the young children under the age of fifteen, because the male priests do not feel it's proper for them to do this menial job. They will also be allowed to clean the church, prepare the flowers and decorate the church as they have always done.

"It was directly after this announcement things started to happen at Radstock Church. First, a very heavy thunderstorm, in the middle of a hot autumn afternoon started and the sky directly above the church turned black. We can go now via a direct live feed to show you what is happening outside Radstock Church."

Mike waited in silence, watching the same film as the viewers were seeing on their televisions. The thunderstorm had been recorded and was shown first and then cut to the live images of the same thunderstorm, with brilliant white and blue lightning striking between the clouds and to the ground. Another man was talking the viewers through the views they were seeing and the weatherman was adding information about the storm and saying it should not be happening.

"The other thing about this thunderstorm, as you can see by the live feed to our radar system, according to our radar images, this storm is not happening. The skies should be clear with no wind or rain. Yet as you can see, the sky is black and rain is hitting the ground so hard, it's bouncing off a good 6cm high. The wind is blowing the trees and leaves in circular motions and it is not the wind which is making the bell toll, it is something else entirely different.

"The fire brigade have put a platform up to the belfry and have made numerous attempts to tie the bell in position and stop it from peeling. This is what happened on the last attempt," the man said and the scene changed again.

A fireman was seen to throw a rope around the bell and pull it back to the steeple. The rope was cut into thousands of small pieces and captured by the wind which took it into the firmament. The fire engine, weighing over six tonnes, was picked up by the wind and dropped into the road 490 metres away. The man in the cradle was also carried by the wind, and everyone watched as he was thrown through the air to the road outside the church, where he was caught again by the wind and placed onto the footpath standing upright and unhurt.

"None of the firemen were injured and this shows how powerful this wind is and that it's not a normal wind. The bell continues to toll, calling for the people to come to the church and hear God's own words and see his power," the reporter was saying.

"Most of the clergy were outside the church when the strange wind appeared inside. It is said that Cannon Mathews, who has just returned from Israel, was carrying ancient scrolls with him that tell of the second coming of Christ. Mr. Hudson, can you tell us any more about this and what you know happened inside Radstock Church earlier today?" Mike asked him.

"Yes, it's true, Cannon Mathews did go to Israel on orders from Archbishop Langley, and he did return with ancient scrolls. During the discussions, some of the scrolls were read and it was said the second coming is going to be on . . ." He paused, feeling very frightened.

"Mr. Hudson, if you know about the second coming and from your own appearance and what has taken place inside the church, you owe to the nation, to the world even to tell us what it is you know. Nobody can harm you here, you are quite safe. Please if you know the date as I'm sure you do, please tell us," Mike said sitting opposite him on the edge of his seat.

The cameraman zoomed in on Mr. Hudson, seeing the thought on his face, the strain he was going through making his decision. He was obviously not lying, by his looks alone people could see he knew something and something had happened to him to make hair turn white in a matter of seconds. Christians and all faiths had a right to know when Christ was going to return to Earth, the question was, would Mr. Hudson reveal it?

There was a long, silent pause, both men looked at each other and the producer held his breathe.

"The second coming is to happen at 06.00 on the 13th of . . ." He paused again wondering if he was doing the right thing, if he was even allowed to tell what he knew. He thought of Mike's words what he had said moments before, was he safe here? Was he safe anywhere from God?

"Please Mr. Hudson, please continue, like I said, no harm can come to you here, take all the time you need."

Time was what they didn't have, the producer had cut into prime time TV and even now the directors were glaring at him from their window upstairs.

"November, its November the 13th in less than . . ." he glanced at his watch noticing the time and making a swift calculation. "In less than 35 hours the Lord will appear to us once again," he said and looked at Mike lowering his head.

It was out and the producer threw his hands into the air, the phones were already ringing with people asking questions. Two minutes of further small talk passed and the directors gave the thumbs up, they were going national and breaking into every channel, including all Sky Networks who were now paying a fortune to be included in this momentous news

A commercial break, slightly longer than normal was next. During those long four minutes, scenery was changed, phones were brought in and a very frightened Mr. Hudson was pushed into the limelight.

"It is said that you saw God Himself, did He speak to you?" Mike asked in a calm manner after the commercial break.

"I don't know where you got that information . . ." Mr. Hudson paused, he was now on national TV and after all, it wasn't his idea in the first place. In a few seconds he decided to make up his own story and become a hero for the entire world. He doubted very much if the people still locked inside the church would ever be seen again, it would be his story and his alone, nobody was with him to say anything else happened.

"Mr. Hudson, you are the first man since Moses to see the face of God and hear him speak, please, tell us, the entire world, tell us what happened in your own words. You owe it to the world," Mike said gently pushing him.

He paused again, "Since you put it like that, it was dark or darkish; there was a very strong wind inside the church. Everything was being blown around, papers, books, even the pews were picked up and thrown against the walls and floor. Candles were flying through the air; flowers which had decorated the church were torn to shreds by the vicious wind as I was being pushed further up the church away from the vestry door."

"What was happening inside the vestry?" Mike asked.

"Archbishop Langley, Dean Huntley, Cannon Mathews, Mrs. Rocc, Mrs. Peters, Mrs Meadow and Mrs. King were all inside. I hammered on the door but it was locked and I could hear terrible screams coming from inside the room. It was horrific as I stood by the door, hammering with all my might and there was nothing I could do as the screams from the people got louder and louder. I don't know what they were going through or what was happening to them. As I looked to the floor, I saw a pool of blood seep from under the door.

"I cried out, shouting with all my might, Stop! Stop it; but the screaming continued and the blood continued to seep beneath the door and past my feet. Then the screaming stopped as abruptly as it started, blood covered the floor by my feet. I stepped over it; I felt if I touched it I would be subjected to whatever happened inside the vestry." Mr. Hudson paused for a moment to catch his breath and do some more thinking.

He looked at Mike, who was sat on the edge of his seat, everyone; including the cameramen were silent, hanging on his every word. He put his hands to his face, as if to hide the tears that were not there. He started to sniff, as if he were crying, the audience which had been quickly moved from another studio to make the interview sound better were all silent and no one moved an inch as they waited for Mr. Hudson to continue.

"A strong gust of wind caught me and thrust me half way along the church; I fell to the ground holding tightly to a pew." That part he knew was true at least, but his mind had been working overtime for the next stage.

"Then I heard a sound. I thought it was the wind at first blowing through the pipe organ but when I looked down the aisle, the organ was torn to shreds. There were tubes, keyboards and parts; even the seat was strewn across the choir area with their pews broken to splinters.

"In the dimness, because it was dark in there, a light started to shine. I thought for a moment someone had got into the church and was about to help me. My heart raced as I called out to let the person know where I was. When I looked towards the light the wind subsided and became a dead calm. It was like being in the eye of a hurricane; the air was still and smelled sweet." His pace quickened as he thrust himself deeper into his story which even he himself now believed.

"I clambered to my feet, holding onto the pew with all my might and as I stood upright I looked into the light. I saw a face, an enormous face which looked like that of an old man, yet he wasn't old. He had white hair and a long white beard, he smiled at me, and I didn't think to take too much notice of other facial features. I was too frightened; I didn't know what was about to happen to me and just looked at the face.

"There was then a sound, like a thousand choir girls, all in perfect harmony which filled the church, or my head. From behind the face, inside the white light, dark shadows flew, actually flew across the back of the head before my eyes. They were passing to and fro, singing this. . ." He paused for a moment trying to find the words to describe his imaginary scene, the press wanted something and he was trying to oblige them; the problem was, he was not good at writing or describing things and making things up didn't come to him naturally.

"The singing was a triumphant volley of angels, praising the Lord God Almighty, who I believed I was in the presence of, it was magnificent. From behind his head the light dimmed to show me what I presumed to be my first view of the other side. . . Heaven itself."

A loud gasp came from the audience as he spoke the words; the cameraman jerked the camera making the picture jump for a moment. Everyone in the world was waiting for him to continue, imagining in their own mind's eye what Mr. Hudson had seen.

"It was the most fantastic, formidable, and loving and caring place I have ever seen; it was green, with field after field of lush green grass with angels dancing and singing everywhere my eyes looked. People like you and I walked hand in hand through the fields and I felt this wonderful feeling of love, peace and quiet, that nothing could harm me, there was no evil at all. It's a perfect world full of love and joy. There is no need, desire or greed; they have everything they could need.

"I could feel there was no fear or hate, I felt that was forbidden, love is the key word in heaven and a most wondrous place to live. If we could but see heaven now, I'm sure we would all commit suicide to go there and be among the love and feeling of being needed. The people I saw looked to be at peace, although they may well be spirits, for they were to me, people in our form. I felt no illness or death lived there as it does on Earth. The scene faded into the white light and the angels were singing again, praising God."

"Did you ask God any questions or did He speak to you?" Mike was asking almost in silence as he too was captivated by Hudson's words.

He believed every word Hudson spoke; he had no reason not to. He was told before Mr. Hudson arrived that he had most definitely seen God with his own eyes and that was why his hair was pure white.

"I uttered a question yes. I asked him, 'Are you God?' He didn't speak like you or I, but a thought entered my head. He replied. . ." He paused again for what seemed hours before continuing."'Yes; he said or thought to me."

Gasps again filled the audience, people stood, hands to their mouths, the producer told the soundman to lower microphone.

"Do you mean he used telepathy?" Mike asked him.

"I suppose you could call it that, he didn't move his lips at the time or open his mouth."

"Then you, in your state of shock could have thought he replied to you and said yes," Mike questioned.

"Oh no, it was a definite thought answer, as I continued to stare into his face, I asked him what he was doing here and why he had demolished the inside of the church."

"What did he say?" Mike asked quickly for the world. He was sure now that he was interviewing the only man to have spoken to God since Moses. He thought he would be famous in his own right and soon be commanding the very best talk shows on TV.

"He said he was making the way for His son, the second coming was nigh and he would not make the same mistakes as He had before. This time, He would not allow His son to be crucified and made a mockery off. If man didn't accept Him for what He was telling them, then He would smite man and woman . . . hard."

"Did he use telepathy again?"

"No, this time he opened his mouth and spoke like I am doing now. He is angry with us for what we did to His son. His words filled the church with a deep voice and his words echoed through the empty rooms. It filled my body, my very soul, as he spoke, a gust of wind left his mouth taking his words with it to me."

"What did you make of his answer?" Mike asked sitting forward on his chair. By now, everyone was captivated by this question and answer syndrome.

"I suppose, but this my own belief that he, Jesus, or whoever will come to us, will not be born as an infant, he will be placed on Earth, fully grown as a man. Either that or he has already been here for many years and has grown up in secret ready to show himself to us."

"Where will this take place?"

"I asked God that question," he sighed and looked into his open hands.

"Did He answer you, were you frightened?" Mike asked and looked directly at Mr. Hudson.

"I was scared like never before, the question sort of slipped out of my mouth. I was too frightened to even think of the consequences, it was like having my first interview for a job, the question just popped out," he held his hands apart, what else could he say? He was trying to think where God would send his son and he didn't have a clue, because he was not allowed into the vestry where the talks were going on and only one man knew the answer, and he was more than likely dead.

"Did He answer you?" Mike repeated, seeing it was taking a lot of effort for Mr. Hudson to speak. "Why don't you have a sip of water?" Mike suggested and indicated the jug of water and two filled clear glasses.

"Oh yes, thank you," he replied picking up the glass and taking three gulps of water. He looked at the glass, realising the water was not what he was used to. He glanced at Mike then at the glass wondering if he had been drugged, then he felt warm and something filled his body and the water was oh . . . . so sweet, he had never tasted water like this before. He was silent, looking at Mike and smiling.

"What is it?" Mike asked.

"Taste it," Mr. Hudson said and pushed his glass into Mike's hand.

Mike looked at the camera then at the audience and the producer was talking into his ear asking what was happening. Mike sipped the water, and . . . smiled.

"God said His son will show himself to us in the church of God, the church in Britain."

"This is God's water isn't it, God is here with us now ladies and gentlemen, here, Tom, come here and take this water and let the audience taste it, only a little sip, I don't know what too much will do for us," Mike replied as Tom came forward from behind a second camera and took the water jug and glass.

Tom sipped the water and stopped dead in his tracks, the water was special he knew and he wanted more of it, much more but he knew he had to let the audience taste it as well; they would know it was not a gimmick.

As Tom passed the water around the front six rows, the jug continued to fill with its God given water, everyone in audience was feeling the love of God enter their bodies and a feeling a well being was rushing through their veins. A few people who had problems with their health, were suddenly realising they were no longer in pain and suffering, the water had cured them of their life long illness, at least it had seemed that long to them.

"Did he say which church?" Mike asked as the audience was drinking the health of God.

"He just said God's church; I would presume He was referring to Westminster Abbey. It is a big church which would hold thousands of people to see and hear his first words to our people. It could also be Radstock Church, because this is where He showed himself to me, so I would say, thinking on it, His son will appear at Radstock Church."

"Did He say anything else?" Mike asked feeling on top of the world. For the past four years, he had suffered with his back after a bad fall and always had a niggling pain in his legs and thighs. He was now realising he didn't have any pain and the only conclusion he could come to was that the water, God's water had cured him.

His producer was giving him the thumbs up signal that everything was going fine.

"God told me I would be spared and see His son for myself; I was suddenly aware I was standing in the air, high above the broken and damaged pews. I was level with the face of God as he breathed over my body and I felt cleansed of all my sins. The wind subsided for a moment and a brilliant white light appeared over the vestry.

"Tiny, what appeared to be stars, twinkled and moved up through the light from the vestry roof. It was I believe the departing souls of the people inside going to Heaven. It stopped as soon as it started and the wind commenced once again. I was lowered gently to the floor and a giant hand of wind appeared and touched my head and a tingling sensation flowed through my body which is when I presume my hair turned white, God gave a final smile and disappeared, and then the wind filled the church again.

"I hurried from the inner sanctum of the church and was blown to the floor by the time I reached the inner oak doors. I felt a presence behind me as if I was being pushed out of the church and no longer wanted there; it was needed for another person.

"As I passed the inner oak doors they slammed behind me and locked. I stumbled to the outer two doors and as soon as I was out of the church, the steel doors closed and locked and the bell commenced to toll." He looked at Mike knowing the last few paragraphs were at least real, even if he had changed them slightly, he assumed he had made a good job of telling the public what they wanted to hear.

"That is an incredible story and I've just been informed the single jug of water has been passed through the two hundred and twenty people and it's still full of God's water. At the moment it is being passed through the crew and the people who work in the gantry. I've also been told that this water," he said holding the glass of water in the air for everyone to see. "This water is making people who have been ill, well; it's as if they have taken some of God's holy water and it really is making everyone feel at peace and in good health.

"I have had three sips of this water and I had a bad back and now I'm cured, at the moment, to prove this to you viewers at home, we are sending for three or four people who are badly disabled to try and prove this before your eyes. So stay tuned for a while longer as we'll be staying on air for the forthcoming time.

"We have had thousands of calls and many people are asking if God is going to destroy the world, can you answer that question for them Mr. Hudson?" Mike asked as he was prompted what to say through his earphone.

"God mentioned nothing about the world coming to an end, although there has been talk associated with the second coming that the world might end. However, if Jesus comes to us again, it must surely be to save us from our sins, not destroy us, or what would be the point in sending him here? God would surely destroy us by fire or flood, or one of the other methods as mentioned in the bible."

"Mr. Hudson, is God a forgiving man, for I have sinned many times, I need to know if he is forgiving," the voice said over the studio speakers from a distant phone call.

"He appeared to be kind, when I was lifted into the air, I felt no pain or the wind, even though the wind was clearly evident in the church," he replied to the caller.

"How will we know the Lord when he comes to us?" another voice asked over the studio phone.

"I'm sure we'll know in our hearts when we meet Him that He is the Lord and the son of God," he answered truthfully.

"Is He likely to go on national TV?" a woman asked next.

"I would imagine Mike will interview Him himself if he gets the chance. I'm sure when He comes to us, someone will get Him onto television, times have changed since He last set foot on this planet. The easiest way to reach the billions of people on our planet is through television, I'm sure the world's press will arrange something for Him."

"Are you certain that Archbishop Langley, Cannon Mathews and the others are dead?" someone asked.

Mr. Hudson paused and looked at Mike, he looked worried and frightened of committing himself to an answer, but inside, he hoped they were. If they were alive, he would be made to look the biggest fool in history and no one would believe Jesus when he finally came.

"From my position and what I could see, please remember I could not see inside the vestry as the door was locked, I couldn't gain access to the interior of the room, I would say they were all with God," he replied choosing his words carefully.

The way he worded his answer, they could be dead or alive and he knew for sure, when he was inside the church with them, they were all with God, so he wasn't lying.

The man on the phone sobbed aloud. "My wife was in there," he uttered and cried. "Why did God kill them?" he continued.

"I'm sure He had a reason for everything He did inside that church, is that Mr. Meadow?" Mr. Hudson asked, thinking he now recognised the voice.

"Yes it is."

"Well, I'm sure God needed everyone in that room for a reason, I did say from my position I thought they were dead. I can say for sure, that they were and are still with God. If they haven't come out of the church, then God still needs them for whatever reason He has chosen. It may be that He has taken them to a higher place to witness the coming of the Lord or to make preparations for His coming. Perhaps He wanted them to see Him in all His Glory and see Jesus before he comes to Earth. Let us not forget, it is said in the bible that there will be witnesses to the second coming and perhaps God has chosen these people to be His Witnesses."

"What do you mean by this?" Mr. Meadow asked.

"It says that God will prepare witnesses to ensure the second coming does not go wrong and the witnesses will be with Jesus when he returns to Earth. The witnesses will then live for a thousand years and oversee Jesus' work on Earth and help Him like his decuples did in His first life on Earth."

"That would mean they are dead right now," he said sobbing over the phone again.

"Mr. Meadow, if they are to be God's witnesses, then they will be returned to us and Mrs. Meadow will be returned to your side, to be with you and love you for the rest of her days."

The questions and answers continued for an hour and no other programme was being transmitted. Finally, Mr. Hudson answered his last question, one he was not expecting.

"If you saw God as you say you did, was God male or female?" the woman asked.

Mr. Hudson was silent so too the audience, which had been quite chatty in the background, now, you could have heard a pin drop. He knew what had transpired in the church and the subject Cannon Mathews wished to bring to everyone's attention. He wondered if this person had seen some of the scrolls, or was even inside the church earlier in the afternoon, she could even be one of the lay preachers or higher up in the priesthood.

"I presume He is male, to me he had a white beard, He showed nothing more to me than his head and a hand. We have always assumed God to be male and I saw a male God; He had the voice of a man, in my humble opinion, by what I saw and heard, I would say God is male, like it says in Genesis, God created man in His own image. By these words alone it tells us that God is male; I don't know where you got this idea from; but God is definitely male," he replied with a grin, then thought better of it and smiled and lowered his face from the camera, praying that what he had said was the truth.

A moment later the studio lights flickered and the power dropped for a full ten seconds. Technicians and engineers checked circuits and output signals, tapes which were recording the show continued to run, despite other equipment failing. The phones were cut and no matter what they tried, they couldn't be reconnected, not even mobile phones worked.

The ground in the studio shuddered as if an earthquake were in progress; people tried to stand as a wind burst through the studio doors into the studio itself. Cameras were tossed into the air as people rushed to the exits, the lights dimmed as John the producer rushed around calling for calm. Suddenly John was caught by the wind and slammed hard against the wall.

As he looked to his left, Mike was next to him and turning to the right, he saw Mr. Hudson pinned to the wall by the wind. As he tried to talk, the wind rushed into his mouth stopping the words coming out, he watched cameras which cost hundreds of thousands of pounds tossed into the air smashed to smithereens; from somewhere deep in his mind he was wondering how he would explain this to their insurance company and who would pay for the broken equipment.

As he looked around, his eyes fell on the monitor secured to the wall on the opposite side of the studio. He could see the three of them pinned to wall by the wind at least a metre from the ground. Then the lights went out completely and a voice filled the studio.

"I have listened with great interest to what has been said this evening; some people cannot stop telling lies, Archbishop Langley and his companions are not dead, they are in my presence in deep discussion and they will remain in the vestry until tomorrow morning, however, I will not allow this fiasco to continue any longer.

"The second coming will take place as you were advised by Mr. Hudson, at Radstock Church. I God, herby invite you to be in attendance with your cameras to show the world the second coming. However be warned, I do not need to use your cameras to show My world of the second coming, I am quite capable of doing this Myself.

"As you can see from your television pictures now, not one of your cameras is working, neither is this scene or my voice being transmitted to the world. For your ears only at this time, I am not a man, your God is female."

The wind stopped and the lights returned to their normal brilliance, as people pulled themselves to their feet, they looked around at the mess. The monitor was still showing the studio and what was happening.

John, Mike and Mr. Hudson fell to the floor with a thunderous bang, drawing attention to their position. As they stood, people who had turned to see what the noise was fell silent as they stared in wonder at the three people standing before them.

"What's the matter with you all?" Mike asked, and then rubbed his throat, wondering why his voice was higher than normal.

"Come on, get these cameras picked up and let's see what's working to give us that picture and cut the outside broadcast," John shouted. Then he too rubbed his throat, something was different with his voice, it was higher than normal.

"We can't stop the pictures, we shut down all the power to the monitor and cameras but the pictures continue to go out," said a voice over the studio sound system.

"What's the matter, why are you staring at me?" Mr. Hudson asked. He too rubbed his throat as his voice was at least three octaves higher than usual, more in the pitch of a woman's voice.

Then the three men looked down; their bodies had changed dramatically. Before the wind and darkness, they were all men, now they were women and their hair was long and pure white, as was their skin. They all had breasts and what lay beneath their new female clothes, they were unsure, but they felt different than they had before and now they were wondering what God had done to them and why She had done it. Their male clothes were gone, replaced by female clothes and high heeled shoes.

"What happened?" Tom, the stage manager asked.

"I . . . I don't know. I felt something touch my hand, then a tingling sensation went through my body, did you hear that voice, a female voice? Mike asked Tom impatiently.

"Yes, it went out on air, the switchboard is jammed with calls."

"I suppose they all think we faked it," John said.

"On the contrary, they are calling to say things happened in their own homes. They witnessed what happened here and know we didn't make it up. Calls are coming in from all around the world," Tom informed them.

Silence filled the studio, when the technicians got the cameras upright, they were amazed to see them in perfect working condition, despite them being thrown around the room and battered against the walls. Everyone in the room had been affected in some way, some men had white hair on their chest, others had their voice changed. Everyone had one thing in common, they all weighed the correct weight for their height and their clothes fitted perfectly.

Two hours later, after thousands of calls to the studio, it was discovered that every person in a radius of a hundred miles of the studio had all lost weight according to their height. All their clothes, no matter how large, fitted them perfectly, also, no matter how dirty or stained the clothes were; they were now like new.

The main change, apart from the three men who had changed sex, happened at twenty hospitals in a one hundred mile radius of the studio. All the patients who were ill, some with cancer, one home with forty patients with AIDS were all cured of their illness. People who had broken bones were healed instantly. People had had suffered very bad burns had their skin healed like new and all their previous illnesses were cured as well, all the people were perfectly healthy with 2020 vision, even twenty five people who had been blind from birth could see, and of the 297 people who were deaf, all could hear perfectly well.

Where accidents happened that evening and people had been killed, they were miraculously brought back to life and were healthy normal people. A blazing fire was extinguished in less than a second and everything put back to new.

Church buildings, some in desperate need of repair looked like brand new. Grass was cut, grave yards put in first class condition, by the end of the night, it was discovered in the area touched by God, not one person was ill. People suffering from terminal illnesses were cured, the blind could see, the deaf could hear, disabled could walk and the insane were now sane. The old, frail and tired were feeling like they were back in their teenage years, with plenty of energy to get through the day; not one person needed glasses, a hearing aid or a stick to help them walk.

In prisons, governors opened their doors as they were told no inmate could ever commit another crime again and God had forgiven them for their misdeeds, even those who had murdered and raped their victims, God had ensured these people had paid their debt to society and the person they hurt was healed, and those people who had been killed, were now in Heaven and their relatives were healed of their grief and forgave the killers. The police stations let all their prisoners go and no one in a thousand miles radius was arrested that night.

People on drugs were clean and people who sold drugs found they no longer had the drugs to sell and their lives had changed, drastically. In a five hundred mile radius of Radstock Church there were no drugs anywhere and as people brought illegal drugs into the area, they were turned to sugar and the people cleaned of their of their addictions.

The roads and motorways were blocked with cars and people trying to get to Churchill and Radstock Church, in the hope the people would be healed.

In every country throughout the world, a small area was picked by God where She let Her miracles work. Since Jesus died on the cross, never before in such abundance, did people bow down to worship God in whatever religion they knew. To them, God was God, and soon, Her next descendent would put everything right. In those small areas where God had spread Her hand, people were healed and made to be the right weight and their clothes fitted perfectly.

For a reason they didn't understand, Mr. Hudson, now Miss. Hudson, Mike now Michele and John now Jenny, waited patiently outside the Radstock Church where it all started. In the graveyard, flowers blossomed and trees were in full bloom with glistening leaves which kept watch over the graves below.

There were still twenty hours to go but these people would be the first to see the new Jesus when He appeared and they prayed He would return them to their previous bodies. There were also cameras by the hundreds and the three people with thousands of others were waiting patiently, in silence, not hungry or thirsty, thinking of what had taken place when God had shown Herself to them.

There was no doubt about it, man had made a terrible mistake, nobody knew how it had happened or why, but they knew it had to be corrected. God was female, that much they knew; She had spoke to them and carried out many miracles, their thoughts were now on the second messiah, would it be male or female?

CHAPTER 5

In the vestry, the discussions continued with the jug of coffee being replenished several times and the tray of sandwiches, being emptied more than thrice.

"If what you say is true, and at this moment in time, I'm not sure what to believe, we have misread the bible for centuries," Archbishop Langley stated.

"I'm sure these scrolls are correct, God is of the female gender. If we go back further, into a dimension where God exists in reality, then perhaps there were many Gods in times gone by; perhaps in reality there are many universes," Canon Mathews stated.

"The universe is immense, you cannot have more than one universe, I do know a little about the subject, my hobby is astronomy and I help teach a young school club the subject. We live in the universe and there are many thousands of galaxies which make up the universe as we call it. The universe is a collection of galaxies; there is nothing greater than the universe, that is the be all and end all of it," Dean Huntley said proudly.

"I don't wish to contradict you, but let me put it to you that our universe exists in Gods own universe within his own dimension. It is said in the fourth dimension you would be able to see all four sides of a cube at any one given time; the fifth dimension is possibly time, what about the sixth, seventh, eighth or ninth dimensions?"

"There could well be any number of dimensions, each occupying the same space and time, separated by a dimension in space, or another universe. They would not interfere with any of the other dimensions or universes. We say Heaven is up and Hell is down, but from space exploration, we know Heaven is not on the Moon. It's not likely to be on any other planet in our solar system or even our galaxy and we have telescopes that can now search light years into deep space and we still have not detected something as large as Heaven.

"We assume Heaven is close to us because it is looking over us and we speak to God in our church. We assume he can hear us and therefore must be in hearing distance, listening on the other side of a curtain which separates our world from his. If we are to believe that when we die our spirit or soul leaves our body and goes to Heaven, then it must pass into another dimension that is only accessible at death. We have never before discussed where Heaven is or what dimension of space it occupies.

"Earth could be in the middle of a million different dimensions or universes, each one existing on the other side of each other and we would never know they existed. This room could be filled with people, this may not even be a church vestry," Cannon Mathews proposed.

"That is purely assumption and has no gravity to it," Dean Huntley said interrupting. "Hypothetical, unproven and dreamt up by people who write science fiction stories, that is all it is, stories. There can be no other universes, and that is fact, God would not allow it."

"I am not talking hypothetically," Cannon Mathews retorted. "If She can see over us, the spirits can see us, then surely, the dimension which they occupy must be a parallel world. They could possibly have a form of translucent one-way mirror so to speak, to separate them from us. I mentioned ghosts before, if they can see our world then why should they not occupy it?"

"Rubbish; you're talking out the back of your head," Dean Huntley interrupted sounding annoyed.

Cannon Mathews paused for a long moment cooling his temper, which was beginning to rise. "The problem is they are not fully in this dimension, but in between. I refer to the tunnel we speak of when experiencing a near death experience. It must be the link between one dimension and the other. In fact . . ." Cannon Mathews paused and rummaged through his briefcase. In his excitement he brought out an old scroll, unfurled it and placed it on the table before him.

"What is it?" Mrs. Meadow asked.

"That is what I was wondering when I saw it, I picked it up not understanding it; neither did anyone else who knew of it or had seen it; but I think Dean Huntley will be able to tell us what it is," he said with an excited grin as if he had found a million pounds.

Dean Huntley sat forward in his chair and studied the scroll rubbing his chin as he did so. He was in deep thought when the scroll turned around by itself. A smile crept over the Dean's face as he clapped his hands together in joy.

"Of course, I didn't see it until you turned it around."

"Nobody touched it, the scroll moved by itself," Mrs. King replied.

"Well, the answer is simple, it's a star chart, of a specific area in fact and here," he stated full of excitement, indicating a large black dot. "Here is a Black Hole. It's an amazing thing a Black Hole, there are many theories about them and this one here," he said touching the dot, "Lies near our solar system."

As he removed his finger, the black dots changed to white dots and were positioned in different places.

"What happened?" Mrs. Peters asked.

"I think I know," Cannon Mathews replied. "When Dean Huntley recognised the Black Hole, he started a sequence on the chart. What we see now is another star system in another universe or dimension. God gave us the opportunity to visit another universe or dimension without having to die, to give us a far better understanding of what space is all about. No doubt there are White Holes in this universe which lead back to our universe and further on to more universes."

"Of course, it was suggested Black Holes are gateways to other universes, but there is no conclusive evidence as we have not yet managed to master space that well. We have only just placed a manned base on the Moon and Mars has just had its first ship land there with sixteen new Martians," Dean Huntley explained.

"So now we understand that different dimensions and universes exist, we can more safely assume that Heaven is in another dimension, reached through a tunnel after we die," Cannon Mathews explained. Everyone agreed silently nodding in the affirmative.

"Talking of tunnels, have you given any thought to our birth?" Mrs. King asked.

"I'm sorry, but I don't follow you," Cannon Mathews replied. "What has that got to do with the subject of space, which is what we are discussing?" he continued, baffled by her question.

She smiled; seemingly looking down on them for their ignorance in Mrs. King's vital link in their new theories.

"Well," she said and paused for a brief moment. "Women carry a baby in our womb for nine months which grows from two small cells. Yet the only way out of the body is through a tunnel inside the woman between the womb and her vagina, then out into the world of reality. It could be at this point when the soul enters the baby and not before; after all, what would a soul want to be inside a woman for nine months if it couldn't do anything?" She was silent for a moment letting the thought sink in.

"Do you know what that would mean?" Mrs. Peters asked.

Mrs. King shook her head; she had just conceived the idea of the tunnel, not the outcome of its importance.

"I do," said Dean Huntley, being more of a scientific mind. He analysed the idea as it was spoken.

"Our thought of spiritual life beginning at conception is wrong; life does exist, but not true life. I have buried many children who have just been born or born a few days and not survived. Doctors dissect them to see why they died, for many there appeared to be no reason in particular; but if what Mrs. Peters said is true, the reason the child dies, is because the true living force, the soul, did not enter the body in time. It didn't traverse the tunnel, the dimension between our two worlds and set in motion everything which is needed for the baby to survive. For the Living Force to survive . . ."

"The main item of a body, the living part of the brain which brings the baby to real life, gives it expression, thinking and reason for being. Without the soul it is just a body, a lifeless body that may live for a few hours then die without the living entity to control it, to make the decision to take its first breath and breathe in oxygen, filling its lungs with fresh air and giving the soul a body to move around in and share the joys of being a real entity in our material world. It may well be able to live here, but without a soul, it does not have the ability to make things, to go to other places and touch and feel and experience places and objects, things it cannot do in its spirit form," Mrs. Rocc intervened.

"Precisely! Therefore if the living force we call the soul is not there, there is nothing for the body to live with, to control it. It doesn't have the determination to live, the willpower to go on and survive; in fact it's a cabbage," Mrs. King concluded.

Dean Huntley was silent, they all were. The idea of the soul entering the baby at the delivery stage had never occurred to anyone before because it had never been investigated or discussed; yet is seemed plausible. After all, what would a living spirit or soul do inside a child for nine months and could two souls occupy the same body at the same time without one soul wishing to take command of the body and fighting for it. Surely if this were to happen, then the new soul, the soul that had come from Heaven, would be full of life and programmed to be hard and fight for its life. It would, without a doubt, fight to survive and fight the soul that was inside the woman. Two souls occupying the same body, even though they were in different parts of it was not possible.

A spirit had to be free, being inside the womb for nine long months it would not be free, it would fight to get out and live the life it was meant to live, long before its due date. It would be restricted inside the womb, despite knowing the mother and hearing things going on around it, it would not be able to interact with any other soul except the soul of the mother, and if it did interact with that soul, the mother would understand she was much more than a woman, much more than a person, she would know and understand as the new arriving soul fought hers, that the body is no more than a biological machine, a carrying implement of the soul which the soul could command and use for its own benefit and it would fight hard to survive and kill the baby inside her to eliminate the other soul. The more they thought the situation through, the more apt it seemed and the more real it appeared. The soul would only enter the baby as it was born and not before. It had another life in another dimension, things to do, places to go and souls to be with and interact with.

The more they discussed the principles in detail, the more convinced they were that it was true. Since nothing happened in the room, they were positive what they had stumbled upon by accident, was in fact reality. Then they questioned if they had stumbled on it or if the idea was put into Mrs. King's mind by God.

Archbishop Langley sat forward and looked at the silent faces around the table before he spoke.

"Having established the birth and death of a spirit or soul and where it goes to in the correct context, we now have a better understanding of our situation here on Earth. However, we can only presume the dimension theory is correct, because we don't have the spaceships to see for ourselves where the Black Holes take us to. Therefore, we will assume this to be correct until God tells us otherwise, and I am certain God is here with us and if She wanted us not to believe this new theory, She would tell us to stop this line of questioning, do we all agree on this?" he asked.

Everyone nodded in the affirmative.

"Who knows, God if He, I apologise She, allows us to continue to live on this planet, She may in Her wisdom, give us the knowledge we need to travel to travel to those different dimensions and universes," he continued.

"I'm pleased to see you're thinking like us at last," said Cannon Mathews, smiling for the first time.

"It's hard to believe in something new, when we have believed that something else was true for so many generations. We have taken it for granted that God is male and He was speaking to man, because we assumed he made man first and woman second, made man in His image. We have lived with this assumption for thousands of years, our new problem is, how do we go about telling the people we were wrong, and how do we convince the Pope and the Archbishop of Canterbury we were wrong. The probability of a girl being the second Messiah will, I am sure, devastate a great many men," Archbishop Langley admitted.

"I agree wholeheartedly," added Dean Huntley.

The women smiled at the Dean and breathed a sigh of relief. Of course the new Christ would have her work cut out, especially winning over the men, who have been ruling the planet since they were placed on it.

"If you have no objection to this question, why here; why England, after all, we are not the greatest nation on Earth and definitely not the most Christian. We have multi-lingual societies now who believe in any number of Gods and different theologies," Mrs. Peters asked, sending everyone back into silence and thought.

"God moves in mysterious ways, this time She has to reach not a small number of people but billions. All who know the name of God and believe in God, no matter what their faith."

"God as a male God," Mrs. Rocc butted in.

"Yes," Cannon Mathews agreed. "But I was going to add, they have split the church. There are many lies told throughout the world, not just by children but by leaders of the church and governments. Our Leaders lie to us over election policies they know as they speak they cannot fulfil. They lie and cheat in business, they lie and cheat in land deals all to make money; money on a large scale, to make them very rich, but what they all forget, no matter how rich or poor you are in life, you enter this world with nothing and leave with nothing. You can't take anything with you except memories, and those are worth more than all the money in the world. To a dead person, a wooden box is a wooden box, no matter what type of wood it's made from, it will not make the body any more comfortable," Cannon Mathews said.

"I don't get the drift of your answer," Mrs. Rocc replied blushing.

"We are expecting a saviour, the second coming will, I'm sure, be more majestic than the first. There will be no Kings or Queens trying to subdue or kill the saviour, at least not in this country.

"In the USA or Russia, things may be different; they may not like someone telling them they are forgiven and the CIA and former KGB would be demanding to interrogate the Messiah and more than likely dissect her to prove she is a living entity of God. In Eastern Europe they might withhold the news and try and stop any mention of the second coming getting out to the rest of the world. They may even try to do this here; we'll have to wait and see what happens.

"In Africa, many places have no electricity or televisions, the Messiah could turn up, speak to millions and the rest of the world would never know it happened. In Europe the police carry guns as they do in the United States. There are too many fanatics who would take it in their head to make a name for themselves just to give it a try and shoot her because they are not Christian.

"It is most likely this country alone, with the police and justice system, the Messiah would at least stand a chance of making the news and telling everyone what she has come here for. She knows the rest of the world would know immediately, wherever a television could be found," Cannon Mathews concluded.

"'So you're saying it's through our justice system, that God has chosen our country?" Mrs. Meadow asked.

"Well that, and also that it's written in the ancient scrolls where and when the Messiah will come?" Cannon Mathews replied, taking the said scroll from his briefcase laying it out on the table for all to read.

A few minutes later everyone sat back in their seats amazed at what they had read. Now they knew for sure where and when the Messiah would come. To their church, in a very short time, but they had, they hoped, worked out the reasons for it themselves, that part they were proud of.

"We should try and inform the outside world, the press should be here to witness her coming," Mrs. King suggested.

"Do you think God will make her daughter sacrifice herself for us?" Mrs. King asked.

"I doubt it very much, we have heard of the miracles of Jesus, and it's not as if we disbelieve Jesus could do those things and we have accepted the miracles even though we were not there at the time. We live in a different world now and we don't crucify people, at least not in this country; perhaps the only way the Messiah could be crucified would be in the courts of Justice," Dean Huntley said solemnly.

"That will not happen, we will not allow it to happen and I'm sure this time, God will not allow it to happen. This time it will be different, it has to be, we live in a much bigger world and have worldwide up to the minute communication. What happens in one country is seen around the world as it takes place," Cannon Mathews stated in a positive voice.

"As we are all agreed that God is female, then the expected Messiah will be female. That the God we presumed to be male could possibly well be the devil and Heaven lies next to us in a different dimension, with the only way in and out of that dimension is through a tunnel of light, either by birth or death. However, once we have spaceships fast enough and big enough, we will have the opportunity, or at least hope we will; to be able to visit other dimensions through Black Holes, what else is there to accomplish here?" Archbishop Langley asked.

"Much!" Cannon Mathews stated sternly "We have to decide whether or not women should be allowed into the church as vicars, bishops and higher up the ladder, even up to the office of archbishop. We also have to inform the public before the Messiah comes when it will happen and where she will appear. We have to inform everyone that God is female, not male, and the expected Messiah will be a woman. We have to make them believe, we their church leaders, have misread the bible in the area of gender. Explain that we have discovered that Hell is here on Earth and we make Hell ourselves. Also, we must declare to the world, and make our position very clear, as to how women stand in our church," Cannon Mathews explained with a voice of authority.

"I thought you mentioned that," said Mrs. King.

"No, I mean before we leave this room, we must agree upon a new form of leadership for the church."

Everyone was silent. "I . . . I don't understand what you mean," Archbishop Langley stuttered.

"Quite simply, we must move with the times. I strongly suggest that we overrule the motion which was made earlier today and allow women right up to archbishop." He looked at his watch seeing it was 06.15 in the morning. "I apologise, yesterday," Cannon Mathews amended. Then he stopped and looked at his watch again, hardly believing how the time had passed so quickly and he didn't feel tired, even being awake for over twenty four hours.

"It is now exactly twenty four hours to the second coming, we only have one day left and so much to accomplish. We have to tell the press where to come first," Cannon Mathews said in an urgent tone.

"It has already been done; Mr. Hudson did that last night on live television. The press are gathered outside our church as we speak," Mrs. Rocc explained.

Everyone looked at her amazed; she had spoken the words, but couldn't possibly know what had traversed the previous evening. Cannon Mathews stood and checked the vestry door which was still securely locked with no key.

"We must assume that what we have just heard is correct, but I doubt if the press are aware of the consequences involved about the gender," Dean Huntley added.

"I now understand what you are getting at," Archbishop Langley said.

Everyone looked at him; he was smiling with a look of pleasure on his face, something no one had seen on him before. He had always been a vigilant man, intent on obeying the word of God. He nodded, but no one was talking. The silence continued for ten minutes as the others watched in wonder, how his face changed while he was being spoken to.

"Cannon Mathews, you are most correct in your findings and suggestions; I have just heard the voice of God, and it was to me, most definitely female. She told me Her daughter will come tomorrow morning with the sound of trumpets and a host of angels."

Everyone was silent with mouths agape. There was a definite change in the archbishop and they were to witness, for the first time in endless years, a man being talked to by God.

"The Archangel Gabriel himself will lead the procession of angels and the daughter of God will be called Mary. She will be met by the new joint heads of the church, King William V and Queen Kate," Archbishop Langley stated. "Mrs. Rocc will be the first archbishop to meet Mary with myself and other disciples of Mary," he continued.

Everyone glanced at Mrs. Rocc who was blushing profusely. "Has he got it right?" Cannon Mathews whispered.

"Yes," she replied in shock.

"Then archbishop, are you saying you will allow women into the church?" Dean Huntley asked, sounding slightly annoyed.

"More than that, they will be allowed and encouraged to move as swiftly as they wish up through the ranks of the church to take every office, including that of archbishop. Mrs. Rocc, the rules of yesterday have been overturned by a much higher office than mine. I have been told to ask if you will take up the position of archbishop with immediate effect. You will bring about the reformation of women into the church and help them up through the ranks," he said and looked in silence at Mrs. Rocc.

"But she knows nothing of the intricacies of the church; by all means allow her to work with you, but she knows nothing at this time," Dean Huntley objected.

"Are you against her being made archbishop?" Cannon Mathews asked glaring at him.

"No, but . . ."

A soft, gold glow was settling over Mrs. Rocc's head; as the glow grew brighter, it slipped down over her body and lifted her from her seat, making her float above the table so she was encased in the light. She was suspended in the air for a full two minutes while everyone was silent watching her. She smiled, almost as if she was being tickled by the light, then she was lowered gently back into her seat and the light gracefully left her body, rising up through the ceiling and out of sight.

"Are you alright?" Archbishop Langley asked urgently.

"I am fine thank you. Cannon Mathews, you speak eight languages, please ask me a question in each language." He did, and she answered fluently in all eight languages, then she spoke Latin perfectly, as if it were her native tongue.

"I fully understand the workings of this church, and every other church Dean Huntley, well beyond your capabilities. I am able to perform any function in any language without having to read from the bible or prayer book. I don't know how, but I know the bible by heart, not just our bible, but new versions of the Old and New Testaments. I know every hymn, Psalm and prayer by heart, I do not mean to boast, but it is what God has given me. I know I must use this gift and it will be needed on the Day of Reckoning."

"The Day of Reckoning?" Dean Huntley questioned. "We have not discussed this day or when we assume it will come."

"It is not far away," Mrs. Rocc answered.

"Then you are more gifted and higher than me. I hereby declare you leader of the Church of England. I bow and kneel before you," Archbishop Langley said and started to stand.
That is not necessary Archbishop Langley, you have a job to do and that is to prepare the way for the coming of Mary. I will stand bedside you and be your equal, as man and women will be from tomorrow forward."

"Equal? I assumed women would demand their right to lead the world where man has been dominant for thousands of generations," Cannon Mathews suggested.

"Man will no longer dominate woman, Mary will see to it. From tomorrow, no man will ever hit or abuse a woman again, if he does, he will be punished by God immediately, and he will not like the consequences."

"I feel stupid asking you this, but did She give you any powers or gifts?" Canon Mathews asked.

"She has given me the power of healing, mind reading and telepathy. For those who will help in this coming event, they will need to use it and most of all, teleportation, so I can travel to any point in the world, where my help is needed."

"You mentioned disciples; do you know who they will be?" Dean Huntley asked.

He looked around the table at the other women, wondering if they too would be included. Then his mind wandered to Jesus, his twelve disciples, all men. He pondered if now, as women were to be higher than men, it would be twelve women, with no men at all being given the chance to represent God.

"God will choose," she whispered and closed her eyes.

The room instantly glowed and a bright light descended over the people in the room.

CHAPTER 6

At midday precisely, the six people in the vestry, who had been talking practically none stop for 22 hours, concluded their work. To their surprise, no one was tired and they had just eaten an excellent three course lunch with coffee and red wine.

Sitting back in their seats they were now sure they had made the right decisions, and knew exactly what they had to do. Saying a prayer, they stood and waited. A loud click unlocked the vestry door and Archbishop Langley stepped forward with Archbishop Rocc by his side. The door opened and Archbishop Rocc stepped first into the church, as they walked to the central aisle, they marvelled at what God had created.

The walls were brilliant white, with a white ceiling and new beams of oak to support the vaulted roof. The aisle was lit with a form of electric candle in a gold holder, suspended in the air from nothing. A gold cross, double the size of their normal cross with rubies, emeralds and diamonds inserted into it, was suspended like the candles over the entrance to the altar. The pews, no longer old and withered, were made of the finest oak and now brand new. The deep red carpet beneath their feet was thick piled and luxurious. As they turned to see the altar itself, a cloth covered the wooden table, brilliant white, inlaid with gold thread. Gold goblets and crosses studded with diamonds lay on the altar and more crosses stood behind the goblets, on small stands.

They each took a new cross and put it around their necks; as they did, they were dressed in new church robes and they were all the right weight for their height. Turning around, with their dress code complete, they looked at the new pulpit.

The pulpit before was a crumbling mass of bath stone, it was replaced with new white marble and a golden eagle made from wood, with two large rubies for its eyes. The large bible, with gold edged paper was placed on its back. Every prayer book was edged with gold and carried a gold inlaid cross on the front cover of the red covered book. Moving to their left they faced the organ, but what a sight met their eyes. The tubes were all new and the keyboard and keys had a sapphire embedded in each one. Stops were made from ivory and the foot pedals looked as if they were made from the finest ivory, not from any animal, but oh, what a glorious sight.

As they turned they saw the giant stained glass windows, with new pictures of God in all Her beauty. Turning to face the rest of the church, flowers of every kind filled the church and at the end of pews; behind them and from their sides came the sound from a choir of angels, their voices soft and beautiful, filled the church, everyone in tune, unlike the choir which had practiced long and hard during the previous evening sessions.

"This is most unbecoming of you God," Archbishop Langley said aloud.

"This is the twenty first century," God replied into his mind. "Now, go through my doors and give my children your news. I will open my doors at the appointed hour tomorrow morning and not before."

They walked slowly through the fragrant church, every flower in full bloom and in their hearts they knew each flower would not wilt, for they were in God's own special house.

When they reached the new oak doors, they opened for them, stepping from the inner sanctum of the church, they entered the outer chamber. Turning around, they looked back into the church, which was now ten times longer than before and four times as wide and filled with pews. Half way up the side walls, were the circle seats, which went right around the church and above this in the gods, another circle of seats and behind them, stairways wide enough for four people standing side by side to take a leisurely walk to the top of the church to hear God's word.

As they entered the outer hall, the new doors closed behind them. The floor, formerly old cobbles was replaced with the finest white marble; the walls were again brilliant white with suspended candles lighting the area. The staircase leading to the upper circle was covered in red carpet and electric candles lit the way, with flowers that brought a warmth and fragrance to the stairs. Before them were the steel doors that stopped people getting into the church. They were still made from steel, but decorated with a picture of Jesus and Mary, who had yet to set foot upon the Earth. Their outlines were in gold, with sapphires for their eyes and behind them, a cross of gold.

Archbishop Rocc turned and looked up; sensing something formidable was behind her. There was the finest cross imaginable, better than anything man had ever created, it was most definitely a woman's design, with diamonds, rubies and emeralds set into the cross, suspended above the doors to the inner church. Upon her gaze, it appeared to lean forward at an angle to face her directly, then return to the vertical position.

"What a sight," she sighed, not tearing her eyes from the cross.

"Truly magnificent," Archbishop Langley answered.

They turned to look up at the clock and then high above them, they heard the sound of a single bell ringing true and clear, filling the church and surrounding town with its vibrations.

"That's a good sounding bell," Dean Huntley said quietly.

"It's even better than that, the church doesn't have a bell, it was taken down ten years ago and the money was never raised to replace it," Cannon Mathews replied.

Just then, the most wonderful peel of bells rang out, each bell in perfect pitch and harmony, yet the church tower only had enough room for one bell, but now it sounded like ten bells peeling at the same time. Cannon Mathews could not hold himself back, he ran up the stairs to the belfry to see for himself what had happened.

He couldn't believe his eyes, before him were ten bells, all shinning brand new, peeling by themselves with no ropes to toll them, this was God's own music, She was calling to Her children. Once again there were suspended electric candles lighting his way and on the belfry walls, were pictures of planets and solar systems on the walls. There was also a special picture of another planet, Utopia Prime, and in the background, there was a huge cylinder, lit up with galaxies suspended inside it.

"When the time is right, I will show you this place and explain everything to you, for it is where you will spend the rest of your life with me in Heaven," God said into his head

"You look different, what's up there?" Archbishop Langley asked when Cannon Mathews returned to the entrance hall.

"A complete set of ten bells and the belfry is magnificent; there are no ropes to peel the bells which are all moved by God. There are also pictures of planets and the most wonderful sight of Heaven, the real sight of Heaven, that no man or woman has ever seen and still be alive to appreciate it."

"We saw it too, apparently we saw what you saw, I just had to ask to be sure. Did God speak to you?" Archbishop Langley asked full of excitement.

"Yes, there is much to do, many people to speak to and we must talk with the press. I just don't know if they will believe our story," Cannon Mathews stated.

"I'm sure they will; have you noticed anything else?" Mrs. Meadow asked, now a full disciple of God.

They stepped forward to be clothed in the purest of white gowns with gold thread running through their garments and shoes of red leather, with gold coloured laces were placed on their feet, yet they were not cold. Stepping back the clothes disappeared, stepping forward they reappeared, it was a miracle.

A moment later a gold coloured light drifted over Archbishop Langley and Dean Huntley, as they were bathed in the light, Archbishop Langley's prayers were answered. When the light left them, they too knew the bible by heart and could speak in every tongue, read minds and as Archbishop Rocc had been given the power of healing, Archbishop Langley had been given the power of healing and understanding so they all had the same gifts. At the slightest touch of his hand, he could make the most stubborn person understand what God wanted and expected of him. He had been given the gift of putting the word of God in every man's heart; he could push the word of God into everyone's heart and be there for Mary.

They all stepped forward and were dressed as God wished them to be, standing before the great doors. They opened in silence with a small gust of wind lifting the leaves into the air. The press watched in shock as the doors, a dull sun bleached red, changed to bright white and gold and opened inwards. The press were not expecting the doors to open until the following morning and had no idea what was about to emerge from within the church. Cameras were switched on and equipment made ready to film whatever they were about to witness.

The group disciples stepped out into the wind, and although the wind blew among them, their white gowns remained at their sides. Not a hair on their heads moved as the doors closed silently behind them, and then everyone could see pictures of angels had adorned the doors. Some of the cameras had managed to pick up the transformation inside the church and many people were talking about it, wondering how it had happened. As the group stepped onto the evenly paved courtyard, a triumphant roar came from the camera crews and news people.

The skies suddenly darkened and lightning struck between the clouds. In full view of everyone and the rolling cameras, God wrote with the lightning into the white stone walls of Her church.

WELCOME TO MY CHURCH

MY CHILDREN

Particles of stone fell from the church walls and those watching knew it would hit and certainly kill the people standing below. As the stone fell, God covered Her disciples with a halo of light and dissolved the brick before it touched their heads. People applauded and a loud cheer erupted through the thickening crowd as word quickly spread something awesome was happening at Radstock Church.

"What happened inside the church?" everyone shouted at them

"Was God in there with you?" another group shouted, trying hard to get their questions answered.

Archbishop Langley raised his hands and a gentle peace fell over the group of people. The new Mrs. Hudson was still being interviewed on television in a small portable studio which had been set up outside the church. She was still answering questions on her made up story getting deeper and deeper into trouble when a message came through to Michelle explaining what was happing outside.

"If I may interrupt you Mrs. Hudson," she said turning to face the cameras and putting a hand to her ear, the producer Jenny repeated her message. "Right thank you, I've just been informed ladies and gentlemen, Mrs. Hudson, that Archbishop Langley and his group are now standing outside the church and the church has dramatically changed its appearance and the people are wearing some form of white robe. We are now changing to the outside cameras so we can witness what is taking place," Michelle, formerly Mike Adams said to the people watching her show.

The lights in the studio dimmed and the cameraman moved around as the view changed to outside Radstock Church. The front of the church and the sound of peeling bells could be heard as the camera panned down to the flattened courtyard.

"This isn't Radstock Church; it's a fake or something. I mean, look at that courtyard, it's flat whereas ours is cobble and has been for 150 years," Mrs. Hudson stated sounding annoyed as any woman could be.

"Also," she continued without pausing. "The bells; Radstock Church doesn't have bells, it only ever had one bell and that was removed over fifteen years ago and we never raised the money to get it repaired and put back in place. This must be a farce," she shouted. Then she heard the bells herself inside a soundproof room.

The camera panned around and video film was showing earlier scenes of what had happened at Radstock Church. Mrs. Hudson couldn't believe her eyes as she saw the people she had said were dead and with God. Now they were standing as real as before outside the church and there in the background, was the most wondrous cross she had ever seen.

"We really are at Radstock Church," Michelle was saying. "What we have witnessed is a true miracle, the Archbishop and his aides are unhurt and alive," she said full of excitement. This was something she had not been used to doing being a man a few hours ago and now letting her feminine emotions run away with her.

Then they showed the falling stone, in slow motion and the words cut into the wall.

"Mrs. Rocc, who previously walked with a limp, now walks upright and said that when God touched her body she was healed. The church is magnificent, as I am talking to you the outside of the church is being altered, becoming brand new, only this is far superior to anything we could build and we are now in a position to show you what is happening," Paul Simon, the outside reporter was telling his viewers.

The cameras showed the church being rebuilt at phenomenal speed and the graveyard looked as if it had just been made. The headstones were being changed to white marble with gold writing looking brand new, with close cut green grass and flowers everywhere. All the trees were in full bloom and apples, large, red and green, were pulling on the branches where birds sang and flew in the trees, proclaiming God's wisdom.

Mrs. Hudson was flabbergasted and shaking in her shoes, being led away to the side of the small studio. Michelle was talking to her, demanding to know what had happened and if she had told the truth earlier.

"It was as I told you," she said, convincing herself that the things said really did happen. Some of them did and now she was trying to convince herself that everything she said happened as well.

"Come on, let's face the music and see if Archbishop Langley can shed some light as to why we have been changed like this," Michelle said angrily.

There were by now over fifty different television cameras outside the church with hundreds of technicians monitoring sound and pictures. The police had their job cut out holding back the crowds which had blocked off roads and made it impossible for traffic to get through.

The three new women walked slowly towards the group who had emerged from the church. Archbishop Langley looked forward and smiled, he held up his hands and the groups of reporters and cameramen moved out of their way. Jerry Andersson, a reporter covering all four major stations just managed to get all the latest information on the new women. He alone was now reporting the facts as they happened.

His voice was low, almost silent, describing in detail the looks of anguish on the faces of the new women as they walked through the widening crowd of reporters wearing dresses and high heel shoes.

"The new women look frightened and tense as they approached the figures that emerged from inside the church. Archbishop Langley and his people have hardly moved from their spots since they emerged, except in Mrs. Hudson's case, she was ejected from the church yesterday. These people walked out and the doors closed majestically behind them and nobody knows what is about to happen.

"There are a few police officers moving around to the group's right, obviously they want to know what has happened inside the church, any damage caused and what exactly was inside. By the look of Archbishop Langley, I don't think the police will get their way.

"The three new women are almost before the Archbishop and have now stopped a few feet away from him, looking silently into his eyes. We have wired them up with microphones so anything they say we will be picked up and we will let you hear it as it happens," Jerry Andersson told the world.

At that moment the sound manager nodded that he was cutting in. "Ladies and gentlemen, we have sound. What you will hear are the actual words spoken by the people," Jerry said finishing his commentary and looking at the monitor.

Archbishop Langley smiled and stepped beside Archbishop Rocc, the side where the police were moving in from. Archbishop Rocc stepped forward and held her hands out to the three women.

"Why? Why did He do this to us?" they asked in unison.

"Because my children, you were transmitting lies about God, when we were hearing the truth. There has been enough lies told the through the generations about God, now is the time for everyone to know the truth."

She looked at their attire and smiled inwardly, God even knew about fashion, she thought to herself. The thought was picked by the other disciples and a chuckle left their lips, showing they were still human.

"God in all Her wisdom has changed your sex and way of dress because you Mrs. Hudson told an outright lie," Archbishop Rocc said boldly.

"What lie?" she asked bewildered, knowing she had told umpteen lies.

"You stated that God was definitely male; unfortunately through the years of translation, the bible was interpreted incorrectly. Cannon Mathews discovered the truth in Israel and we have discussed this with God."

"Then God was inside the church?" Michelle asked nervously.

"Oh yes, God was with us all the time."

"Mrs. Hudson said the church was extensively damaged inside, there are many expensive artefacts and gold crosses said to be worth a fortune in there," Jenny stated, being a reporter to the bitter end.

"Stop a moment," Michelle said suddenly. "You said God is feminine, did I hear you right?"

"Yes I did," Archbishop Rocc replied. "God really is a woman, we have talked in great depth and read scrolls no one has ever seen before, it's true, God is of the female gender. We will not go into detail right here at this precise moment, but when everything is prepared, we will explain to everyone where we made the mistakes. This is where the programme is being transmitted from, did you know that in deepest Africa, where there is no electricity, everyone there is seeing this as pictures in the air, the screen is a kilometre high and two kilometres wide and it is being broadcast in every language known to man?" she asked them.

"You jest lady, no one can see pictures in the air, let alone place them there. How can you put a screen that big in the air, it would weigh . . . tonnes, the glass alone, and how are you powering it?" Michelle asked smugly.

"Behold, this is exactly what the African tribes are seeing at this very moment. Of course they are hearing our words in their tongue," Archbishop Rocc replied.

She held up her hands and before everyone's eyes, were four of the largest screens that anyone had ever seen, two kilometres long, one kilometre high, all floating in the sky, so everyone could see what was happening, no matter where they were.

"This is impossible yet . . . it's really happening; I can see the screens, I can hear the people, hear us speaking and yet the sound isn't loud. How are you doing this?" Michelle gasped.

"This is nothing compared to what God can do and it's God's wish that the entire world will see at the same time what will happen when her daughter arrives tomorrow morning. Those who are asleep have been woken just to see this, to witness what has happened so far." She smiled and looked at the coloured policeman making his way forward through the crowd.

"Come here my child, you are not of this country, but a tribe in Africa; please tell the ladies and gentlemen what is being said," she asked him in Swahili.

"It is amazing, truly amazing; I can understand it better in my own language than in English. Yes I can understand what I am saying as I speak to you in English and I am hearing my words in my native tongue, this is truly amazing," he replied in his native tongue.

"He couldn't even see the pictures ," Michelle stated arrogantly.

"The screens are not flat, they just appear flat to us, no matter where you are in relation to the screen, it will appear to be flat, and you will see the entire screen. Would you walk to the church wall and look at the screen and then walk back again to us, and see the power of God in our technological day," Archbishop Rocc said inviting them to see for themselves.

"Look at the density of the crowd, you'll never get through," Jenny said nervously.

She didn't like parading around as a woman in a dress; too many men were staring at her, looking at her breasts and observing the way her hips swayed back and forth, the way her long hair swayed as she moved her head, it was hard being a women. She had never before thought of women this way, the way they felt being stared at all the time, the way men glared at them, the way he would have looked at a women, the things which went through his mind, how he could caress her, show her a good time until he had finished with her.

Now she realised what it felt like, how some women didn't wish to be leered at. She didn't want to be mauled by a man, any man, especially her old self, because she knew what she was capable of and how her old self treated women, made fun of them, made them look small so he could look good, feel big and important. Now the shoe was on the other foot, and the four inch high heels were hard to get used to, yet God for some reason, was forcing her to wear them, like all the female clothes she was wearing right now, he felt strangled by the tight fitting undergarments, how could women feel good in them she couldn't understand, but right now, she had no option but wear them.

Archbishop Langley waved his hands and people moved out of the way in his indicated direction forming a path to the far gate which was now clear, with no one hurt in the move.

"How did you do that?" Jenny asked him.

"Just go and see for yourself, see the screen and come back and tell me what you see," Archbishop Rocc stated.

He left, taking a camera crew with them, seeing for themselves what the screen looked like from another view point. It was true, no matter what angle you looked at the screen, it appeared to be flat, there were no wires, aerials or gantries to hold the screens in place. They just hung there, three hundred metres in the air, powered by an unknown force. The electronics inside must be fantastic, Jenny was thinking to herself, but she was losing the thought of reason, why was she was sent to look at the screens? They were not man or woman made, they were made by an entity that was God, and God could manipulate matter in any way She liked. As far as she knew, the screens were a figment of her imagination; God was putting the pictures directly into her head. It was just better for her to see them on a screen, something she was used to seeing, to stop the fear and realise that God was here, making things happen. If God really wanted to, She could do so much, even make this world disappear with everyone on it; gone in the blink of an eye, never having existed.

When they returned they were lost for words, Archbishop Rocc smiled as the group gathered around her.

"We feel you are not very happy with your new attire of clothes and sexuality," Archbishop Rocc said smiling.

"We have wives and children at home, we are not lesbians; we are men or were men with male feelings. What is to happen to us now, do we have to dress like this for ever more until we die?" Michelle asked.

"Women wear these clothes all the time; they are glared at all the time and treated like you like to treat women all the time; but you know what men do don't you, because you used to do this all the time to women. How does it feel being on the other side, being glared at, the thoughts of what they could or want to do to you; how they would like to treat you, do to you, whether you like it or not. Does it seem fair, would you like me to bring a man here and show you how much of a woman you really are, show you how it feels to raped?" Archbishop Rocc asked.

"No, I'll change I promise," Michelle and Jenny said together.

"Your sex was changed so that you would know how it felt for God. She doesn't like being called a man or referred to as a male, no more than you like being a female. Each gender has its own tastes; you learn to live with the sex you were given at birth. Unfortunately there are some that would love to be changed like you have been; I refer to the transvestites and gay communities.

"Man will have a deeper understanding of their needs and tomorrow, a lot of lives will change forever," Archbishop Rocc explained.

"What about us?" Michelle asked.

Archbishop Rocc raised her hands and prayed silently, then, laying her hands on each woman's head, a soft white glow covered them. Everyone watched on the big air screens what was happening; a minute later the glow disappeared and the women were now men, dressed in their former clothes, but now with a greater understanding of how women felt and wanted to be treated. They also had a far greater understanding of God and what Her church was for and why it was needed. Never again would men treat women like they had and they would from this day forward, have a much greater respect for the church and God.

Michael, God has chosen you to be one of Her disciples; you will go forth and spread the name of God, knowing Her to be female and being cleansed of all your sins," Archbishop Rocc stated before everyone.

"A disciple of God?" questioned Michael, his shortened name of Mike he was used to no longer adequate.

"We are disciples of God, Her new disciples as in the beginning; people will need to know firsthand, and understand what God is all about. Have no fear; you will not start your work until the second coming has happened."

Archbishop Rocc placed her hand over Michael and he was covered in a gold mist which stayed with him for several minutes. When the mist left him, he knew what God and the second coming was all about and could speak every language known to man. Mary would very soon step foot upon the Earth and the truth would be told to the people of the world, whether they wanted to hear it or not; they would hear it and understand it.

Archbishop Langley stepped forward and raised his hands calling for silence. When everyone was quiet, he started to speak.

"My fellow parishioners everywhere and councillors of all the churches of the world, I speak to you now in your native tongue, direct into your heads so you know that it is I who speak to you.

"When we were in the vestry of Radstock Church with God, She showed us where we have gone wrong and how we have misinterpreted the bible. This will start with us calling God a female, rather than male and we will also change the way we have treated our women who wish to enter the church and preach to the people in the name of God. The vote which was taken yesterday to cast women from the church has now been changed, by a much higher member of the church than I or the King of England. God Herself has ordered that women be allowed into the church and shall rise to the highest order and take the lead in our church.

"With this being said, Mrs. Rocc, formerly a Lay Preacher, has been ordained into the church by God Herself, she has been blessed with the gift of being able to speak every language in the world and many other languages which we shall soon come to use, and all of this will be explained when Mary comes to us tomorrow. As of now, Mrs. Jean Rocc will be known as Archbishop Jean Rocc and will be above me in the church."

He turned to face Mrs. Rocc and smiled at her. "Please kneel," he said quietly, but everyone could hear every word he said.

When she had knelt, he blessed her before everyone and named her Archbishop Rocc, when she stood, he knelt before her and kissed the large ruby ring on her right hand which had just appeared. He then stood and kissed her each side of her cheek.

"I know give to you, your people Archbishop Rocc, may you forever serve them well and speak only good of our new God. May God bless you and protect you for all time."

Making the sign of the cross, he turned and blessed everyone before him; with the short ceremony completed, everyone applauded and cheered their new Archbishop, welcoming her to the new church of God.

CHAPTER 7

Police Chief Inspector Adams had had enough of the fiasco he thought was happening at Radstock Church and ordered his men to move in. The crowd was large and they were small in numbers, the sooner they got the group of disciples or whatever they were in their custody, the sooner they would get off duty and go home. If this charade lasted any longer, they would be here all night, as the police moved forward; they commenced to circle the group disciples.

"Chief Inspector Adams, is this what you intend to do when God sends Her daughter into this world; arrest her for what, being the daughter of God?" Archbishop Langley asked in a loud, powerful voice.

Archbishop Langley held his hands in the air and the officers stopped in their tracks, no matter how hard they tried to move forward, their legs would not move, only Chief Inspector Adams was allowed to move, slowly in a controlled manner.

"I don't know who you think you are, but . . ."

"I am Archbishop Langley; that you know to be true for we have met on numerous occasions. You know me and all the people here, except of course in their new roles. We are all disciples of God and God told us who we would be, we did not proclaim ourselves to be disciples of God. Surely you will not persecute us as the Romans persecuted Jesus and his disciples in their day? This is not a barbaric country, however before you speak; the world must know a little more of the meeting which we held with God.

"It was decided after much discussion," he said to everyone, raising his voice slightly, we will allow women into the church up to the highest rank as I said a few moments before. The first woman archbishop, Archbishop Rocc, will be known as the Archbishop of Churchill; this small town will be put on the map and talked about through all time. Women of all ranks will be asked to move as high as they wish, as swiftly as they wish up through the ranks of the church."

A loud cheer erupted from the crowd for a second time from women gathered at the rear of the congregation. Many of them had remained there since the previous day, wondering what was happening inside the church and not believing a word that Mr. or Mrs. Hudson was saying. They never believed that God would kill Her chosen people.

"We, that is man, got it all wrong; the Garden of Eden was a testing ground for the species known as man. It was put in space to see if man could live without his companion, woman. To see if a man would dominate his woman or treat her as his equal and woman proved to have a mind of her own and tricked man into eating the apple which kicked them out of the Garden of Eden. It was Eve, not Adam that was put in the Garden of Eden first.

"The main thing we have discovered is that in the beginning there was a Mr. and Mrs. God, not a single entity, but two of them. Heaven is in another dimension and lies right next to us and is separated by a curtain, how we get here and there is through being born and dying. Hell is ruled by the male God, the person we call the Devil and Hell is here on Earth as near as we can determine.

"We have been so wrong over many things and there are many things we have to put right. The spirit, or soul within us is what makes us work, without it, we will die, perish. The soul enters our body a few minutes before we are born; it is therefore conceivable we were very wrong about the abortion laws which must be changed immediately, although I am sure when Mary comes to us as a fully grown woman tomorrow, she will change women's views accordingly and the men will have no option but to hear them and do as they are told, or fall by the wayside. No longer will women be governed by men and told when they can or cannot have an abortion just because the man wants what he thinks is his. The child was never his in the beginning; the child was always God's.

"Space exploration will be given to us and we will continue to move forward and explore space to understand what was always meant to be ours for the seeing, not the taking. That is God's universe out there, and we were meant to see it, sooner rather than later. It is where we will see many of God's true wonders and we will be able to travel to different dimensions and see different galaxies through what we call Black Holes. We will be allowed to travel deep and fast through this dimension and others, except of course, the dimension God has chosen for Herself.

"There is only one entrance to that dimension and that is through our death. God has shown many things to us and other people," Archbishop Langley paused for a moment as he received a message from God.

"I have just been informed that scientists from both sexes have been given all the information they need to build spaceships that will take us far past the speed of light into the depths of space. The people of Earth are very soon going to see new frontiers," Archbishop Langley explained.

"And you my friend are coming with me," Chief Inspector Adams said snapping a set of handcuffs onto Archbishop Langley's wrists. He grinned when his task was done.

"We are no longer ordinary people, we have been touched by the hand of God and now I will show how we have been changed," Archbishop Langley said smiling.

He disappeared and the metal handcuffs fell to the ground with a loud clatter. Archbishop Langley materialised two metres behind Chief Inspector Adams. There was a gasp from the audience then laughter as Chief Inspector Adams turned aimlessly looking for him.

"I am behind thee," Archbishop Langley said aloud.

The crowd laughed as Chief Inspector Adams turned to see his captive facing him smiling and perfectly calm.

"What are you worried about?" Archbishop Langley asked.

"For a start, what happened to the inside of the church, to all the artefacts and gold crosses? I know they haven't been removed because the security vans are still here."

"They are all safe in God's house, no one will take them."

"What about the destruction, we have to see for ourselves what has happened and how much it will cost to repair the inside of the church. Someone will have to be held responsible for the actions and destruction within the church. It is a crime scene and must be treated as such until we clear it," Chief Inspector Adams said confidently.

"Is making good a church a crime, is cleansing people of illness, death or creating miracles a crime?"

"If it is not done in accordance within the law it is," Chief Inspector Adams replied sarcastically.

"Don't speak rubbish try and understand that God alone is the law; She makes the laws and is responsible for punishing those who disobey her laws," Archbishop Langley replied.

"I think you've had a knock on the head," he replied angrily.

There was a flash of lightning and a loud clap of thunder, so loud people were forced to cover their ears. Lightning filled the sky; striking from cloud to cloud, clouds to the ground where it struck cars, trees, bringing them down and catching them on fire. Huge holes were burned into the roads and pavements as lightning struck with such force it destroyed the concrete. Suddenly, a brilliant bolt of lightning came down just before Chief Inspector Adams.

An image formed, ghostly pale, almost transparent and Chief Inspector Adams was forced to stare at it. His eyes could not close and he could not raise his arm to hide the apparition from his glare. He froze to the spot, his legs unable to move.

"You dare to question one of my disciples? From this moment forward I will rule this planet and see to it that no harm comes to any of my daughters of Earth. My children who disobey my commands will feel my wrath as if punished by their father's belt," a commanding female voice stated and could be heard by others around him.

Chief Inspector Adams trembled in his shoes, feeling the breath of God over his body.

"From this moment forward, you will stop drinking alcohol and refrain from smoking and taking any form of sociable drugs, especially the cocaine you steal from your own safe when it should have been disposed off. I know all the secrets your officers try to hide; how they sell the cocaine and other drugs to people in the street which they take from drug offenders. I also know how you take what you want from recovered stolen items which should be returned to the people they were stolen from. You say you are the police, but some of you are far worse than many of the people you lock away. Your lungs and kidneys are in a terrible state and you will die early if you continue this barbaric form of socialising. I will cleanse your body and mind for any further desire for these misbegotten things and you will in future never desire to partake in them again, including taking things that are not yours."

The image blew its blue breath not just over Chief Inspector Adams, but everyone around him and instantly everyone felt much better and all their ills were cured. Chief Inspector Adams slipped his hand into his right jacket pocket and handed the half used packet of cigarettes to the ghostly figure before him. They instantly disappeared as did everyone's cigarettes and drinks they had on them. In all the local shops that stocked these items, they too disappeared and in their place were bottles of pure clean water and good food, apples, oranges, all forms of fruit which were perfectly formed and not a blemish on anything. People were going to the shops as if pulled by the bottled water and fruit.

The shopkeepers took a small amount of money and for those who could not afford it, filled their bags with food and water and they left without paying. The shop keepers didn't mind, because as soon as one shelf emptied, it was filled again with different forms of fruit and water.

The Ghostly image stirred and rose into the air, as if looking about at the people gathered around it. Once again it blew its blue breath over the crowd and everyone rejoiced in song of *Amen!* Then it lowered itself to look directly at Chief Inspector Adams.

"I will choose you as my Chief Defender for my daughter's arrival. You will be in charge and keep the peace, although I doubt this will be necessary. You will explain the second coming and what is needed to be done to all the police officers in your force. You have no need to worry about the state of My church; it is in excellent shape, just like the outside, at no expense to the people of this planet.

"From this day forward, I will ensure all churches of all dominions will be under My care, I alone will be responsible for their upkeep. Never again will anyone need to give a coin to the collection plate. The ministers will be kept by Me, the poor and destitute in other parts of the world are already feeling my love, and fresh food and Holy water is passing through their veins. No longer will the gun crazed thugs be in charge of my children, their weapons have been taken from them and disposed off. Food now grows in desserts and barren land where millions of my children once lay dying and starving in their mother's arms. I know, for I have been watching, seeing many countries working hard to help feed these people and give them clean, fresh water, while others leave them to die or take away the aid that has been sent to sell it to the highest bidder or use it themselves. I have punished these people already and I am now taking over and have opened the land and wells, filled them with pure clean water, made the land able to bear crops and the crops are ready to pick and will continue to be so for many years to come as the people rise and take their children to join other nations of this planet equally.

"Take heed of what My disciples say; they have more powers than any person on this planet and know how to use them for good. They can travel anywhere in a moment of time by just thinking of where they wish to be. They each know their own jobs and what is expected of them. It will not be like before when My disciples had to walk for many days and nights to get somewhere, now it is instantaneous, just like their power, and I have complete faith in them."

The ghostly apparition dissolved with a brilliant flash of lightning and a loud clap of thunder. Everyone surrounding Chief Inspector Adams was now silent and when they returned to their senses, those who drank or smoked, found they no longer desired to do so and when they looked into their pockets for a packet of cigarettes to throw away, they discovered they were already gone.

God was most definitely moving in mysterious ways; everyone understood what mistakes had been made in writing and understanding the bible. They also realised they had misused their lives, treated others unfairly, upset their own families, broken laws of man and God, but that was changing; very fast.

"While we are here," Archbishop Rocc said. "I would like a meeting of all high church members in two hours at Westminster Abbey," she continued watching herself on the large TV monitor.

"That is a long way, I will give you a police escort to get you there," Chief Inspector Adams replied.

"We will not require transport like that; we now have other ways to get there. However, you could ensure that other church members get there on time," Archbishop Langley said calmly.

"Your word is my command," he replied and left to make the arrangements.

The three men returned from womanhood were very pleased to be male again; however, they wondered now if they had made the correct choice: especially with the way the world was starting to turn.

"Is there anything we can do?" Michael asked.

"You can ensure all your equipment will be ready for tomorrow. Ensure your managers keep all the airtime clear so people around the world can see what happens here," Cannon Mathews explained.

To everyone's surprise, the group vanished and appeared a moment later inside Westminster Abbey. Already bishops and ministers were beginning to assemble and they all appeared shocked, very shocked when the group of ministers appeared by the altar. Bishop Harding rose and shouted as the group made their way to the front of the church.

"This is blasphemy, blasphemy I say; women being allowed into the church; it's against everything in the bible, God being a woman be dammed. You are all fooled; this is not the work of God, but the work of the devil I tell you; it is lies you have been told, this jumping from one place to another is the devil's work not God's. If God had wanted this, then he would have put a man in control not a female, the lesser of our species and the first of our species to take a bite from the forbidden fruit.

"I will not agree to women moving into the church, it is wrong and against everything in the bible," he shouted aloud, waving his bible angrily in one hand high in the air.

Suddenly his bible glowed and turned to dust in his hand, a new bible, in a beautiful gold bound green cover filled both his hands. He was forced to sit and open the first page. Then his eyes read the second, third and fourth pages; getting faster and faster until his eyes moved at a blur and his mind absorbed the full contents of the new book. He recalled every word on every page and knew the new bible, written by the hand of God from cover to cover.

Twenty minutes later with the church now full, he sat upright looking about the church as if woken from a dream.

"Are you alright?" an usher standing next to him asked.

Bishop Harding looked up from his new bible and tapped the cover gently with his right hand.

"Oh yes, perfectly fine, now; I have never felt better in my entire life; do you know; we have all made a terrible mistake?" he sighed and looked at the people gathered inside the abbey.

When everyone was seated, Archbishop Langley and Archbishop Rocc stood before their brothers and sisters. To their amazement, Bishop Harding stood, moved into the main aisle and walked slowly but purposely up the long aisle, looking at various vicars, cannons, bishops and female members of the church as he passed them. The two archbishops were expecting trouble, but they knew God would protect them from the words he might say. When Bishop Harding reached the beginning of the altar aisle, he stopped, standing before the two archbishops. He bowed his head low and kissed each person's back of the hand; archbishop Rocc first, not because she was a woman, but because he had read something which had changed his way of thought.

Standing upright, he smiled and turned to face the congregation before him. Silence filled the giant abbey, as everyone wondered what was about to happen. This was most unexpected as it was Bishop Harding who had led the revolt against women entering the church, yet here he was kissing the woman's hand.

Before the three people stood a multitude of microphones, far from the days of shouting to get your voice heard at the very back of the abbey, it was now a simply quiet talk. Bishop Harding tapped the nearest microphone to ensure it was working, looking silently about him. The two archbishops had allowed him to have his say to get it over and done with before they commenced their work.

"Fellow brothers and sisters of the church, when I entered this abbey today I must admit I was annoyed and angry at what was about to be suggested to us. I was sure in my own mind this was the work of Satin; that somehow He had managed to worm his way into the House of God.

"I was even discussing the matter in hand with other brothers, giving our sisters a dirty look, for which I have greatly sinned. No matter what we think of other people, we should never go to the extent of trying to wish them out of our lives.

"I held my bible tightly in my hand ready to quote verses from God to you. I was going to explain how Archbishop Langley had been taken in by the Devil, and then something happened to me, or rather my bible.

"Before my very eyes and others gathered here in this abbey, it turned to dust in my hands and then this new bible formed in my hand. I don't know how or why, but I sat down and read the bible from cover to cover in a short twenty minutes and I can recall the entire book as if I had been reading it for fifty years, yet it took me just twenty minutes."

A ripple of voices filled the abbey in wonder and disbelief.

"I know this hard to believe, but it did happen, the bible I read is the new bible, a present from God I thought could not exist. Yet I can recall every word, sentence, verse and page, as if I were reading it now and I know I will never forget its contents.

"In this new bible it tells us everything, where God existed in the beginning, how the Earth was created, the galaxy and universe. It was indeed all created by God, but God also created other dimensions for us to see and visit many other wonders in this universe which is much, much larger than we ever dreamed.

"I will allow Archbishop Rocc to explain in detail what transpired, however, what you must understand, and this is going to come as a shock to a lot of you, I am dropping my case against women entering the church; in fact, I am all for it; they deserve the church more than we do. In my time as bishop, in every service I have taken, there are more women in the church or abbey than men. We always have more women in our choirs, men rarely join in unless they are forced or dragged along to a service by their wife or partner or they happen to be sweet on a girl in the choir. Through all time women have been more church minded than men and I urge all of my followers to stop the motion against women entering the church immediately, if you do not, you will be forced to do so by a much higher power than any archbishop on Earth and you will not be given a choice, it will just happen and you will go along with it.

"One last thing, I fully accept my new archbishop as head of the church; she was chosen by God to do this job and she needs all our support if she is too succeed."

He turned, looking at the surprised people standing behind him who smiled in silence, bowed their heads and watched him retake his seat.

Archbishop Langley stepped forward with Archbishop Rocc by his side; behind them were Cannon Mathews, Mrs. King Mrs. Meadow and Dean Huntley.

"Fellow brothers and sisters," Archbishop Langley started. "Something very strange and beautiful happened to us a short time ago in Radstock Church. We were in the presence of God like never before. We have been shown, beyond all possible doubt; God is of the female gender." He paused for his words to sink into everyone as a great gasp filled the abbey.

Despite all their security, some of the television studios had managed to get cameras into the rear of the church and they were showing everything live to their worldwide audience. In deepest Africa and other continents where it was impossible to have power, God had supplied television screens, huge screens, ten meters wide and high floating in the air, and people watched in amazement as the scenes unfolded before them in their own language.

"It is not our fault, for we did not know the bible was incorrectly translated, although there have been many people who have said certain parts were wrong. So it happens a great part of the bible is incorrect, God sent a man to do a special job; that was Jesus and he was crucified by man; this time She is sending Her daughter to explain God's wishes and desires to us; her name will be Mary.

"She will come to us as a woman in a few hours time at Radstock Church. I personally have apologised to God for our misunderstanding on behalf of the Church of England and all deities everywhere. It is time for the people of this planet to unite and praise God on high for all Her wondrous things She has done.

"Women are to be allowed into the church without any further delay. Those people who are awaiting inauguration, it will be done so this very day; any woman; and by now there will be many who have well surpassed us in our beliefs and true understanding, will be appointed to their just position.

"No doubt God has been talking to many women this day, telling them in Her special way, blessing them with powers no mortals possess, making them into disciples. Long ago in the beginning of Christianity; Jesus chose twelve just men; this time there will be many more just women and men.

"Our world is much larger than it was two thousand years ago and many more people live on our planet and thus will be harder to contact. We are about to enter a new world, a new timeline and we have a new start to our lives.

"I will now hand you over to the first lady archbishop, who from this moment forward will be known as the First Lady of the New Church. This will be a position much higher than my own and even higher than King William V."

The abbey was silent as First Lady Rocc stepped up to the microphone. She spoke with ease and understanding, fellowship and kindness, yet her voice cut deep into the assembled souls sitting before her.

"Sisters and brothers, I have been given this position by God, I did not ask for it. My body has been cleansed of all Earthly ills and my soul has been touched by God's own thoughts. I have accepted this position, not because it was on offer, as a high position within the church, but because God wished me to accept it on behalf of Her daughter Mary.

"We gathered here have been blessed by God, we all have powers we did not ask for, we have jobs to do and we will carry out Her work. You will hear us speak to you in your native tongue which is just one of the gifts God has given us. We need you to back us, not to move against us. When Mary arrives She will tell us what She wishes to be done, and Her words will be the words of God. I know many of you will be frightened of what is to come, especially as we have misread the bible. This will not be the end for us, God is not about to destroy us, if She would have intended to do this, She would not be sending Her daughter to us very soon. We are now entering the dawn of a brand new era, which will change our lives forever. This planet alone, out of all the millions in the universe, has been chosen for us to colonise.

"However, this is not the only planet inhabited by life forms which breathe and walk about the surface; there are many inhabited planets throughout our galaxy and universe. The other news I can confirm to you is that there is more than one universe; we are one of many universes which lie in the multiverse.

"It was stupid of us to disbelieve the astronomers and physicists that said there is bound to be life out there in the universe, I can say truthfully that this galaxy is swarming with life forms like us and we will very soon go out and meet them and they will come here to see Mary. It is now our job, to join with other species and spread the word of God. This will be like before when Jesus came into our midst; this time however, we will not be spreading the word to Romans and other non believers, but to different life forms that inhabit our galaxy and the multiverse. There are already many species who believe in God, but God is sending Her only daughter to us on Earth. Our purpose now, is to build a better universe and for all humankind to go into space and see the wonders God has made for us to see. We will pass on Her name and tell others about God and what She wants them to do and build a better universe for God's children in which to live.

"Mary will help those who understand space, have a much deeper understanding of it. Those of us who understand and design spaceships, people She has given this special gift to, will be taken to a place where they will put their understanding to good use and build us enormous fast ships that will take us through space to distant planets and stars. Those of us who remain on Earth will live in peace and harmony and we will prepare this planet for others to come here and see the true wonders of God and how She has changed Her world. This planet will be the head of the church of the universe and we will travel far and wide to spread the word of God.

"As we move out through the Cosmos, Her disciples will help heal people who are sick and spread the word of God until every species in the entire multiverse hears of Her name. Our planet will be led by women and men together in peace and harmony. There will be no more fighting, no squabbling for small pieces of land; it will all change.

"There will be no more petty crime; anything will be dealt with harshly by a disciple. There will be no more prisons; every prison gate will be opened tomorrow and the prisoners will be set free, they will be returned to their homes if they have any and those that have nowhere to go will be given new places to stay. These people are changed, and deeply sorry for the crimes they have committed. If they did not repent their sins to God, then they are no longer with us."

John Alderson, a vicar who was against women entering the church, stood up tall and spoke to First Lady Rocc.

"If God is so powerful, why can't He or She do the job itself," he asked boldly, and looked about at the people sat close to him and nodded at them, thinking he was doing well. Those that looked up at him glared with half closed eyes thinking he had gone mad or had fallen asleep and missed everything First Lady Rocc had just said.

Total silence filled the church as First Lady Rocc steeped from away the microphone and took the three short steps down onto the main red carpeted aisle. Eyes followed her as she walked slowly and purposely up the aisle towards a quivering John Alderson.

She looked at him kindly, almost pitying him that he would, or could not understand what had just happened and what was about to happen.

"John," she paused and touched his trembling hand. Everyone was now silent, wondering if First Lady Rocc would turn John to a pillar of salt or something just as horrific.

"If God did everything for us, there would be no purpose to life and life would not be worth living. It would be a waste of time, toiling to gather knowledge and understanding why we are here. If we can understand why we are here, then we will appreciate what we have. You don't appreciate anything until it's gone, then you realise what you had and want to have it back. If God were to take away all the tress, we would miss them; if God were to take away all women on this planet and put them somewhere else, man would miss them and go looking for them. They would build spaceships and try to find them and bring them back to be with them; if they didn't, man would become extinct. We need each other to live here and continue our existence; do you recall the dinosaurs; I'm sure you do, but we all miss them and some people are even trying to bring them back, despite their enormous size and how dangerous they can be, we would say they have the right to live. Do you understand what I am saying, we need to be seen to do things, to examine everything around us, to explore and discover the answers to questions we do not have yet. God put us here for a reason, and it was not just to please Her or say I did that. This is why we are here and God put us here to find the answers to questions. She did not put us here to give us the answers; we are to find the answers, when we look up at the stars, we can see they are there and how beautiful they are, what we need to do is build rockets to get there and see their wondrous beauty close up. Do you now understand what I am saying to you?" she asked calmly.

He moved slightly, in pain from an old wound in his back which had never healed properly. He had put up with the pain since the 2003 war in Iraq when he was shot in the back as he helped two soldiers who had also been shot. He was a vicar then in the army and due to his injuries, had been disabled out to return to being a vicar in civilian life.

She saw his pain and rested her hand on his head, saying a silent prayer and concentrating on John. Heat filled his body, a strange healing heat went right through to his bones; he could feel the heat leaving her hands and in shock, remained silent as the healing heat mended his body, filling him with the glory of God.

When she lifted her hand, the heat dissipated immediately and he twisted his new body, feeling old aching areas of his back and legs renewed, the pain now completely gone, he felt renewed, regenerated, like a new man.

"I know exactly what you mean and say. I am now healed from my pain and the suffering has been taken away from me. Thank you and thank God for this; I sincerely apologise for asking such a ridiculous question in the first instance and for being against women joining the church. I hereby, before everyone here, renounce my former post of Vice President to the Council of Women not entering the church. I also declare that I am all for women to enter the church and will go as far to say, that First Lady Rocc is a credit to women all over the world and may God give her a long and healthy life and protect her," John replied, and took both her hands in his and kissed the back of both her hands before standing and bowing his head in respect.

Everyone was amazed at his sudden change of heart and the look of freedom on his face, made him appear ten years younger.

"Before we go any further, are there any ministers here who are suffering in any form of pain, no matter how small?" First Lady Rocc asked. She knew the answer immediately, there were eighty three people in the church who were suffering from long term pain and sixty two who were suffering from short term pain and varying degrees of headaches, arthritis and general aches and pains.

"All those people, do not be afraid, just make your way to the end of your pews and kneel in the aisle," she commanded.

She waited a few moments for the people to move and knowing some were being stubborn or embarrassed to move, commanded every person who needed her help to move to the aisle. She then walked up and down the aisle, placing her hands on their heads for a few seconds healing them. Not only did she heal them, but filled their souls with the love of God.

When the healing was done, she looked hard at the people gathered before her and spoke again to the congregation and the few television cameras at the back of the church.

"You have always believed in God as being in the image of man, yet in one way He is. In the beginning there was also a male God; like so many couples today, they had a difference of opinion and went their separate ways. When God's do this there is or was friction. Mrs God created a perfect universe and the Earth, the male God tried his best to upset matters and that included man. So He falsified things and gave the idea to man that He had created everything when it was in fact His wife.

"Who would think a woman could create anything as grand and gigantic as the universe? It was put to man that a woman could not create such beauty; she had no imagination for this. However, the female God did create this; and man was falsely led to believe that everything the woman had done was created by a male God. In our time, women build houses, create things of great beauty, understand the internal combustion engine and have a full understanding of particle physics. Yes my brothers and sisters, women can do everything man can do and do the job just as well." She paused for a long moment for the whispers to die down.

"The list is endless, women understand the integrate parts of nuclear power stations, our gas and oil fields and even go into space or deep down beneath the sea. We now sit beside man, not behind him. In many fields we are ahead of man but we do lack some of their strength. However, there are some women who have larger muscles than many men, they can do equal and heavier jobs with the way they have built their body and work hard to be man's equal in the workplace. Some women have broken into the field of sport where a hundred years ago they would have been laughed at for even suggesting they run or throw a javelin."

There was a trickle of laughter through the congregation and First Lady Rocc looked up in astonishment.

CHAPTER 8

A man wearing a long purple robe entered the abbey looking around at the congregation. He saw the splendour of the abbey's interior and looked up to the high rafters as if for the first time, taking in the size and what the abbey meant to people and started to laugh. He knew the people had come to listen to First Lady Rocc and what she said God was telling her. He too had a purpose and knew he would have his say in this, God's church in due time.

He looked just like any other member of the clergy, except . . . there was something odd about him and the nature of the way he walked slowly, but boldly to the end of the aisle. As he passed the pews, slowly making his way forward; people felt a cold wind blow over them, sending a shiver up and down their spines. The lights dimmed and some lights went out completely. As he passed people, they became aware he was not what they thought he was and was no ordinary man. For a short time people wondered if he had been touched by the hand of God like First Lady Rocc, but as he passed them, they knew otherwise.

The vibrations were not the same as First Lady Rocc and her assembled disciples, this man was a stranger to them and the church, even though he looked as if he belonged here. He appeared to have part ownership of the church, and the way he moved along the aisle, he moved as if it were his property. As he approached the altar, First Lady Rocc knew he would have his say no matter what she said or did. As he got closer, she felt his aura and knew instantly this was not just an ordinary man; this person was something far more horrific.

He stopped ten pews away from the first steps which led up to where First Lady Rocc stood. They both stood, looking at each other for a moment and First Lady Rocc fell silent, waiting for the person to say something.

The people at the front who had not seen the intruder or been touched by his inflictions wondered what was going on and knew something was not right. As they turned to look back, they saw the intruder and suddenly felt their skin begin to crawl with fear. The look on First Lady Rocc's face told everyone all was not well within the congregation; something was about to happen and they were in the thick of it.

She stood her ground, looking straight into the eyes of the stranger who had come among them, his eyes, even from his distance were different. They were pure black, which appeared to glow and have a depth about them, drawing light into them which it would not let go. First Lady Rocc suddenly recalled what a Black Hole was and now she looked at the person before her, she realised his eyes were just like two Black Holes. She tried to use the gift God had given her to look into the man's mind, but felt a mental block so powerful, her thoughts were rebuked instantly.

She gave a sudden gasp, realising who this man, who dared to enter their abbey was. The rest of the disciples were silent, bound to the places where they stood, mouths closed, minds closed to everything that was about to happen before them. There was nothing they could do to help their leader, they were paralysed, not with fear, but because they were told not to move a muscle by a mind much more powerful than anything they had ever known or encountered before.

The man looked up, the hood of his cloak fell back to reveal a well groomed, casual, short hair cut. His hair was so black, never before had anyone ever seen anything as black as to swallow light, it looked as if it went into eternity, having depth and volume within itself. He took five more steps and stopped.

People were now staring at him, wondering who this fellow could be for their minds had not conceived the ultimate answer. They were on the eve of meeting their beloved Christ, woman or man, they didn't care which, they had waited patiently for 2000 years, and nothing could intervene in God's wishes. Nothing at all, or could it?

Some men at the back of the abbey strained in their seats to move and get a glimpse of the man before them, standing at the front of the abbey facing First Lady Rocc. They had come to accept that God was female, Bishop Harding had proved it beyond all doubt and First Lady Rocc had healed people before their eyes. The scientists were wrong, there may well have been a Big Bang, but God had made that Big Bang happen.

Before them was living proof in First Lady Rocc and her disciples, that God had spoken to them and told them the entire story of the true beginning. Very soon, God's daughter would be among them, if they lived long enough; the man standing at the front of the abbey was no other than a scoundrel, obviously having been touched somehow by the hand of God and taking the power into his own hands was trying to overthrow First Lady Rocc and take the Earth by storm.

Then they recalled the bible, what had been foretold about the future. Bishop Harding flipped through the pages of his new bible which he held in his hand. There, he was reading word for word, before his very eyes, was the fable of what was to happen. Written thousands of years ago, he read it time and time again, he now knew the answer. The problem facing them was their belief; not just in God, but in the people standing before them, strong enough for the future to survive.

Bishop Harding doubted if he had been blessed with any power except being able to read very fast and recall the bible at phenomenal speed. God was with him before; he prayed silently She was with them now. God had to be, without Her, they would fail this . . . test! A test of their belief in God Almighty.

As if to show off, or indeed because he grew tired of walking, the man wearing the purple robe rose a few centimetres from the red carpet and moved forward; his feet not touching the ground. He was gliding forward like some form of magician, as the group before him stood their ground. Behind the stranger the abbey lights were being turned down from their former brilliance.

Nobody noticed the dark creeping up on them, as they could all see just as well what was happening. Everyone wondered how far the stranger would get before something happened. Would First Lady Rocc be indeed the all powerful First Lady she was made up to be? Bishops and Cannons alike sat petrified in their seats, unable to move their feet. They could not even shuffle along the pews; they only had movement of their head and arms alone. They could not even speak to their neighbour. All they could do was look forward and pray to God She would get them out of this unbelievable situation.

Cannon Mathews noticed the stranger as soon as he entered the abbey and knew who it was and how he got past the security people outside the building. He had read about this very moment a week ago in Israel; he wondered at the tale back then; but now it was unfolding before his eyes.

Nobody else knew what was about to happen until he was standing before them; by then, it was too late to escape. The door to their left would take them to the safety of the outside world, in here; they were at this man's mercy. Canon Mathews knew he would show no mercy, especially to them.

Not turning his head, Archbishop Langley thought at the giant gold cross at the rear of the abbey. It was made from the finest gold and was very heavy; it had also been blessed many times by various minsters, bishops and archbishops through the centuries. This was the only thing that could help them in their hour of need.

Under his perfect control, the cross moved forward, slowly at first getting higher into the air by itself, towering above them and over them until it came to rest, suspended in the air over First Lady Rocc.

Everyone was amazed at what had transpired and a single loud breathe filled the abbey. Archbishop Langley could feel the cross being forced back against his will, but he kept his belief and thoughts firmly on the cross over them.

"So you have you reared yourself from the depths of damnation," First Lady Rocc stated in a firm, loud, voice which was heard throughout the abbey.

The man smiled, turning his head to his side, lowering himself to the ground. "You know who I am then?"

"I was expecting you; however, I did not envisage your audacity to enter the House of God."

"Ha, ha, ha, ha, ha! The House of God! This is my Father's house and I will not consider it otherwise."

"Have you a name by which I may address you?" First Lady Rocc asked calmly.

"A name? Any name you wish you may call me but for those imbeciles who dare turn from the truth, call me Mark."

His voice suddenly changed, sounding like a thunderclap; it filled the abbey with vibrations of power from one end to the other; from its bowels to the very top of its spire. The bells shook and cracked with its sonic vibrations and despair filled the faces of the people in the congregation.

"Mark! Mark!" he repeated, his voice filling every nick and cranny of the abbey. In some places particles of masonry fell from the walls to the floor below as his voice went deep into the thick stone.

First Lady Rocc lifted her hands high, and then brought them down outstretched to face her oppressor.

"There is no need to destroy that which is not yours by right; you are Mark . . . The Anti-Christ, am I correct?"

At that moment, four jet black ravens; as black as Mark's hair flew above him and landed on his outstretched arms, two each side. They were large, much larger than any Earth-like raven. They glared at First Lady Rocc as if ready on command to peck her eyes and face to shreds.

"Do you fear us so much that you need to bring your familiars with you?" First Lady Rocc asked.

"You have an entire congregation at your side, or don't they believe in you any longer?" he retorted in rage then laughed haughtily.

"Oh they believe alright, they believe in the truth; they are willing to accept it even though it is not what they expected. What do you intend to do with them; make them kneel before you so you may lead them into damnation; Hell itself?" she replied raising her voice, filling the abbey matching that of Mark's voice octave for octave.

"Very clever," he roared. "I will stop you all from leaving this planet and travelling to other worlds. My Father has curbed you in the past, it will continue to be so in the future," he ordered.

"We have made a great error, but it was not our fault. We worshipped God as a male for thousands of years and your Father loved it; we have been worshiping a man who didn't create anything of which he stands for. Your Father, the Devil, call him what you will, did nothing, yet he has never come forward to explain anything to us. He has allowed us to praise Him because we believed it was a male God who created this planet, the universe and man in his image. Man is his own image, but something this beautiful could never have been created by a man; he does not have the understanding for such beauty as this planet and the universe.

"He can build great cities, cathedrals and abbeys of this nature, but in all his creating, he doesn't know how to put love, tenderness and feelings into it. A house is empty without special things a woman adds, a woman's touch created this planet, nothing else."

"Rubbish!" Mark retorted and waved his arms in rage. The Ravens flew into the air and perched on a high beam overhead, looking menacingly down at the people below. They were there to protect the Anti-Christ and protect him they would from anything that dare lay a hand upon his holy body.

"It was written many years ago you would come to us before the second coming; I'm surprised you left it so late."

I have been among you for a long time, Cannon Mathews remembers me I'm sure," he laughed and made an image of an aeroplane.

First Lady Rocc did not need to turn her head to see the look on Cannon Mathew's face. She read his mind and instantly knew Mark had tried on every occasion possible to stop Cannon Mathews' retuning to them with the documents which told the truth.

"But he still got through despite all your efforts," First Lady Rocc stated.

"By pure chance I was called away and missed him on his last connection."

"You messed up your assignment, God alone helped get Her messenger through; it was Her who called you away. It was Her who sped the train and cycle on its journey; despite all your interventions."

"Do not try and ridicule me woman, I have more power than you can ever imagine," he threatened.

The floating cross had not escaped his gaze and he began to wonder how she accomplished such a fate. As he scanned her mind, which was hard to do as she put up so many blocks, he realised it was not of her doing. He pointed a finger at the cross, expecting it fall to the ground and crush the group, such was the size and weight of the gold and diamond encrusted object.

The disciples concentrated with Archbishop Langley and the cross remained firmly in place. Mark tried a second and third time to lower it but failed to move it from the disciples grasp. Lowering his hand, he pointed his finger to the pulpit. A ball of fire hit the pulpit from nowhere and was instantly put out by First Lady Rocc's single thought. She was learning fast how her God given powers really worked, what they would do. Now she realised why they had been given these powers; God envisaged this happening and how Satin would send His son Mark to test their strength before Mary came to Earth.

A second bolt of fire headed for the end of the altar and a third, but in mid air, they were extinguished.

"What! How dare you!" Mark hollered in frenzy.

"Whatever you attempt to damage in this or any other church of the true God, you will meet resistance."

"Huh! What are you going to do about your beloved congregation and the reporters at the end of this abbey?" he asked smugly. "If I so wish, they will remain in this state or any other state I command them to be in until I deem it they to be released. You will crawl to my feet and beg my Father for forgiveness; you will pray to Him and beg Him to allow you to live in His Kingdom."

He glared at the cross which was slowly brought forward. For a minute second, First Lady Rocc glanced at the cross, then added her thoughts to the those of Archbishop Langley and the disciples and sent it on its way; sailing on an invisible wind down one side of the abbey up to the upper gantries and back down the other side. It passed the group of disciples and continued its path around the organ, choir stalls and back to the front of the abbey.

Mark followed its route in silence, seeing the clergy smile and able to move once again as light filled the abbey making it brighter than ever before.

"I think it's time for you to leave; ask your Father to stop this fight and join His wife, it's never too late to make amends. To say one is sorry for what has gone wrong and put those things to right."

"I will leave when I am good and ready and not before. Never! NEVER on your request or order," he bellowed. His voice was instantly calmed as the cross moved forward and lowered; hovering a short distance from his head.

The Ravens disappeared with the passing of the cross, which now moved slowly forward and Mak looked at it with anger on his face and in his eyes. He did not like being threatened, especially by a cross of gold he could not talk to. As it got closer to him he backed slowly away with the cross continuing its silent advance. Mark turned and lifted himself a few centimetres off the ground then commenced to glide along the aisle, occasionally looking back over his shoulder, cursing First Lady Rocc for showing him up in the abbey he assumed his Father owned and controlled. The four Ravens left their place of hiding high up in the abbey, flying ahead of him as the great doors opened. Once outside, Mark and the Ravens instantly disappeared.

Stunned clergy watched the cross continue its journey to the main entrance and travel outside into the crowd. There it paused, as if scouring the area before it to ensure Mark and his familiars had gone completely, before returning unaided back to its resting place above the altar.

"The abbey is cleaned of all evil which was once here," First Lady Rocc told everyone. "Peace and fellowship be among you and remain with you for the rest of your days."

"Who was that?" a timid voice asked from the back of the abbey. A small boy covered in dirt and grime, clutching a wooden crutch, stumbled forward as if prompted by some unforeseen force.

His small voice filled the abbey with question after question; questions that would like to be asked by many clergymen who sat in the congregation; all wanting to know the answers.

"Come forward my child, take the weight from your crutch, come to me," First Lady Rocc said softly, beckoning the boy forward. He walked at a faster pace, despite his broken leg which had never healed properly.

As he got closer to First Lady Rocc, he stood more and more upright. He dropped the crutch some way back and was now hobbling, much better than before. A few more steps and he stood upright, straight and proud, walking like any other ten year old boy on two good legs. His skin, clean, his clothes clean, his soul cleansed of all sins a small child could possess. He stood in new clothes, his old clothes transformed by a miracle of God before First Lady Rocc.

"Who was it?" he whispered.

She smiled down at his innocent face and gently placed her hand on his head, blessing him before the entire congregation. How he got into the abbey was of no concern; the fact he was here was desired by God.

"The man who has just left was the person we all feared would come. He is called, Mark, the Anti-Christ and it is written in the new bible that he would come unto us when Mary, the daughter of God is about to enter our lives. Mark is the son of the devil as call Him. We now know he is real, a true God in his own right and was at one time, married very happily to his wife, the God we pray to and as yet, we have no name for her, except to call her God Almighty. For whatever reasons, it seems they parted, like many husbands and wives do today and went their separate ways. The female God created everything we see here today, the Earth, planets, stars and universes, the male God became jealous and tricked us into believing it was He who made all this and that He is the one true God.

"By instilling His will of power and corruption into men's minds, He has tried His best to stop the true creator finishing what She had started. The male of the species, ruined a perfectly good experimental ground, The Garden of Eden.

"The Anti-Christ is among us and we must all tread very carefully; the second Messiah is about to enter our world and she will be at Mark's full mercy without our belief and trust in her. The more people Mark can turn against her, the harder it will be for Mary to bring total peace to our planet and send woman and man on our way to the stars to do God's holy work in the biggest way we have ever dreamed," First Lady Rocc said to the boy and the members of the congregation.

Silence filled the abbey as everyone clearly understood what their role would be. In exactly twelve hours, Mary would set foot upon the Earth, the planet created by her Mother, destined to be the starting block to spread the word of God. A stumbling block was already in place, trying to block her way, Mark. The path would not be easy; life in the past had proved that.

One thing First Lady Rocc and her fellow disciples knew, this time God was on their side. She would do all She could to help them, but in the end, it was up to woman and man to sort out their world, a world which had once been controlled by man and the devil, ever since it was taken from God.

CHAPTER 9

As the congregation settled once again, with the eviction of Mark behind them, everyone understood how the devil was trying very hard to mar Mary's special day. He would go to any lengths to disprove the true God was female. To stop the truth being told about Him and what He had done when the universe was created; how He had taken the Earth away from the woman He once loved.

In the television studios, John the producer was suddenly faced with a dilemma. A man had entered the studio unannounced, demanding prime time like Mr. Hudson had been granted. He had something to say to the people of Earth; he needed to warn them about something and he was becoming very persuasive.

As John looked into his eyes, he thought he saw four jet black ravens emerge into the studio behind him but knew it was impossible. Never before had he stared at a person so impolitely for such a long time.

"I suppose you could, we are getting ready to cover the arrival of the second Messiah at Radstock Church, I'm afraid the only programme showing after 5.50 tomorrow morning, except for the children's programmes on separate channels, will be the commentary on the second coming. Every channel is trying to cover it, and fighting for space at the church, which is very small. You won't find another station to accommodate you," John explained in a flat emotionless voice.

He knew something was wrong with him, but he was not sure what. He didn't feel . . . right for some reason. He wasn't himself, he had the feeling of being controlled by an outsider, he was definitely not himself.

"I am aware of that and that is why I have to talk to the people of this planet before you do this, or least the people of this country. Can you go worldwide, right now?" he asked.

"Why should I allow you talk to the people, who are you, are you somebody important, because I don't recognise you from anywhere?" he asked, trying to form the words in his mind. Then he sniggered, which turned into a loud laugh then he stopped himself, placing his hand before his mouth, pressing as hard as he could to try and stop himself laughing.

This he knew was not him, he never laughed before people he didn't know, especially clients who could have a lot of backing, but once again he felt he was not in control of his senses.

"I am Mark, the Anti-Christ as First Lady Rocc calls me," he replied honestly.

His answer went deep into John's very soul; it filled his body, making him quiver in fear and dread. John turned to face his Managing Editor Ian Sandeman who was talking with the floor manager Steve. Steve did all the background work and checked everything was going on schedule.

"Ian, Steve," he called gaining their attention. The two men turned and looked at John then made their way towards him, as if being called by the King of England. To them, John was indeed their King.

"Yes boss, what's the problem?" Ian asked gaily.

"This is Mark, he is the . . ." he was about to explain who this man was but wondered if they would believe him and decided against it. "Mark has something very important to say about Mary, it would be very good for public relations to allow him some viewing time. Not only that, he knows a lot about her and we will have the only access to him, no other channel knows about this. We will be miles ahead of everyone else.

"Set the studio in dark blue with cream chairs," he glanced at Mark. "No, make those dark red chairs. I want a film showing different parts of the world, especially Africa, where the starving people are. I also want another tape showing what man has achieved and gained in material means; you have ten minutes," he ordered and glanced at his watch.

"I'll get onto records, but it will take a little longer than ten minutes to compile the tapes you want. You know I need at least twelve hours notice," Steve said annoyed at the short time he had been given for a job like this.

"Here," said John, handing him a ten terabit USB Scandisk. He had no idea where it came from or how it came to be in his possession. Steve took it and looked at the label, then whistled at the size of the unit.

"Where on Earth did you get this from?" he asked stunned.

"Don't ask questions, take it to VT Editing and get them to make it ready to play."

"It will work on your devises I can assure you," Mark added.

"I've never seen a scan disc so small holding that amount of memory, where did it come from?"

"Would you like more?" Mark asked.

"What? You're kidding me surely?" Steve replied amazed.

"Here, take these," Mark said and handed him thirty of the small devices all a thousand terabytes each.

Steve looked at what he had in his hands and knew his boat had come in. He could sell just one to Microsoft for a million easy.

Mark was reading his mind, seeing the want for money and power fill his soul. Steve was about to walk off when Mark stopped him and took out a small piece of paper and handed it to Steve and another to John and Ian.

"Shall we say payment in kindness for giving me your air time, as long as I want though?"

"As long as you want? I thought it would only be. . ." John looked at the slip of paper which was a cheque and on it was his name and the payment of ten million pounds made out to cash.

"Take all the time you like," John said, "is that right boys?" he said, asking his two colleagues.

They looked at their cheques made out for the same amount.

"You bet!" they said together smiling, thinking of what they could buy and what they could do.

The colour of the stage was blue, which could then be superimposed with special lighting to be any colour they wished. The chairs and curtains were of the same colour and could also be turned into whatever colour they liked to go with the scenery of their production. This helped when a guest was changed over in the commercial break and when the programme came back on; it was as if they were in a different studio.

John was confused again for a moment as to where the Scandisk and cheque came from, but he soon dismissed the idea and thought of the money and what he could do with it once the night was over.

Steve counted them in a few minutes later, the music started and the studio lights came up. The view was immaculate, much better than anything they had ever done before. The colours appeared to be so true, as if they were outside in natural light, something no computer programme could ever accomplish.

The scene playing behind them on the large screen was of Africa, with starving children and adults in desolate deserts hot and clingy. Vast expanses of desert, shifting sands and nomad tribes surviving off what they could find; the heat from the sand could clearly be seen rising in the air through the midday sun.

The children appeared again, this time in vast camps where they starved to death, their fat bellies telling everyone they were not fat through food or drink, this was malnutrition, the children were dying. Even Ian rubbed his collar with his finger feeling the heat coming from the scene before him. It was just as if they too were in Africa with the children. Whatever was happening was not natural he knew, but he was not about to stop it and watch his future disappear, for he was also sure, if he stopped the programme, the Scandisks would disappear and so too would the cheque. Whatever was said or happened, would have to go out, this was his payment for getting the job done.

"Good evening ladies and gentlemen and welcome to our studio audience. We have with us tonight a man who we shall just call Mark. He has been listening with great interest to the stories about the second coming and only feels it proper; the public should be made aware of a few basic facts before Mary, who could indeed be an imposter, appears to us tomorrow morning."

He couldn't believe what he was saying after what had happened earlier in Westminster Abbey. He believed in Mary, yet here he was putting the woman down and she had not had the chance to utter a word to anyone. He started to ask himself how Mark could know more than First Lady Rocc. He was certain it was only her and her disciples who had talked with God. Mark was definitely not a disciples; he was not even in the church when everything happened.

He paused in his thoughts for a long moment as his mouth kept talking about Mark, telling the world what a good man he was. He turned the question he was about to ask over and over in his head and Ian was looking at him making threatening signs with his hands behind the camera which brought him back to reality.

He turned to face Mark. "Mark, you have prepared a special film for us, what are you trying to tell everyone?" John asked and sat back in his chair, still feeling the heat from the camp showing behind him.

They turned to face the screen behind them and the television camera zoomed in on Mark and the picture, showing them on a split screen.

"The scene we are seeing is what it's currently like in Africa; it clearly shows the poverty and neglect of world leaders. For the price of a pay load of nuclear rockets on a nuclear submarine, these people could sustain themselves for an entire year."

"A proper hospital could be built, hospitals and a health service like they have never seen before in their country. They could have schools, tools and machines to help grow and plan their future and enough food to keep them going all their lives. It is not only the world leaders who are at fault here, God in all Her wisdom, if She really exists and loves Her children as the bible leads us to believe, could change this desert and the ways of these innocent children from certain death, to a life of fun and health."

"I know it's late, but hasn't God just accomplished what you have just said? Putting that which was wrong, right?" John asked in a professional manner.

"Did we have to wait two thousand years?" Mark retorted. "This could have been achieved years ago and saved millions of lives. All it took was a single wipe of Her hand; she could have changed this desert into fertile pastures which would produce food for millions of people who have done nothing untoward God, when they arrived there.

"I ask you, if God is all loving and kind, why does She allow this type of poverty to continue? Why does She allow wars, diseases which kill us like AIDS and Cancer; heart problems, strokes, and then force the governments of the world into a state of war so money destined for health care, cannot be spent in that area?

"You would think this to be a peaceful planet, a planet where folk could live life to the full with enjoyment and good health. God is so say sending Her daughter Mary here tomorrow to put right the wrong doings of man. Yet God could have achieved this Herself so many decades ago; why wait until now?

"If you really think about this deep down; if your son or daughter was being crucified on a cross, would you allow him or her to suffer? Suffer so say for the good of mankind?" Mark shouted, getting his words across and heard.

"Surely he did it to save man's soul?" John replied.

"But did he die for us? What do we know of his time, I mean really know about Jesus? I'll tell you what we know. . . Nothing! Absolutely nothing!"

"What about the words in the . . ." John was interrupted in his question which he did not like. Mark was running away with his show and he was starting to take offence at it, despite the money. Then a thought entered his head.

"The amount on the cheque has just been doubled; allow me to continue in my way." John looked stunned and allowed Mark to continue.

"There are no permanent records we can decipher with the utmost accuracy. People from the church now say the bible was translated incorrectly, that is what they are saying, I heard it myself today. The church leaders are now changing the words in the bible to suit themselves, or that which they wish us to follow and believe happened to better their image.

"Will Mary really appear tomorrow morning or, and I am asking you this very seriously to consider my proposal; could it be the first generation of super humans? People who have powers more than the average man or woman. I talk here about telepathy, telekinesis, teleportation, mind reading; being able to alter molecular structures, alter the way molecules move, alter forces of gravity so things can move on their own accord, or appear to. Mind controlling, so they can control what you think and say. If enough of these super humans got together, there would be no end to their power and what they could make us believe. These people exist and in the most unusual places in the world; so can we assume that the woman who tomorrow will call herself Mary, the daughter of God, is no more than a normal, well, perhaps not normal; but super-normal, super human female with the power of her mind at her disposal. What I am saying is; will Mary be a big hoax?"

He paused, making everyone in the studio and at home really think deep down about his question. There had been hoaxes in the past; another would not be out of the question. It was possible with today's micro computers and with the help of mind power; there were only a few people who disbelieved it.

John suddenly realised what he was saying, despite all they had witnessed that day, Mark was putting a question mark over the morning's procedures, and he was making people doubt God. Not only God, but Her son and daughter; John wanted to stop the show here and now but felt helpless to accomplish it.

He tried another set of questions but they all came out wrong, no matter what he said came out just like the script before him, only he could not recall having a script before. This was a live show, it was a talk show so there was never a script, yet here it was on the teleprompter word for word.

"If we continue, we move to this country, America, showing the giant factories and offices; now what does this tell us about God here?" John asked Mark.

The scene changed showing cars, banks, stereo systems and television screens which occupied entire walls called wall screens, a luxury of the early 2020's. The pictures behind him moved on, showing more material possessions which man had made for himself out of greed. Credit cards to show wealth and houses, larger and larger; occupied by small families with staff to wait on them hand and foot.

"This shows us," Mark started to explain. "How man alone has achieved his position in society today. It is not through the help of God, who I very much doubt exists. If God had kept to Her bible she would not have allowed to man to prosper so well in some countries and do so bad in others. If God really is there and is a woman," he paused and chuckled arrogantly. "Then no doubt She would have looked after her daughters on Earth and brought them forward in life. She would have made woman the prime subject on this world, not man. Woman would rule man, not the other way around. If I were God; I would ensure my true like species ruled, but not a corrupt man who I put on the Earth for woman's mate.

"It would be woman who had the majority of wealth and they would hold onto it. I ask you John; if you were in charge of this world, this entire planet; would you let a species unequal to that of your own rule you? Would you allow other people or even monkeys or apes to rule the people you had placed on the planet if they resembled you?"

John paused not sure how to answer with what happened the last time he persecuted God.

"No, I would not, but there again, I don't sleep with monkeys or any other animal for that matter but I do see your point. Women are slowly becoming more equal with us I agree."

"I ask you this; can you honestly imagine a woman, no matter how intelligent, creating this planet and the universe? She wouldn't know where to start," Mark sniggered, showing men had all the answers and could solve all the problems.

He was managing to put women down to the pits and disgrace God no matter what sex. Moreover, he was doing his best and almost achieving his goal, to prove to the people of Earth there was no God. That man had achieved everything and through science alone he was proving man would have been created regardless, even over the millions of years of evolution which it is said it took.

The universe was created by the Big Bang; it was correct and almost proved beyond all doubt. Where was God, especially the female God when all this took place? Mark didn't have to say those words; he had managed to put them into everyone's mind.

People were ringing the studio, giving their points of view, putting questions forward they would like Mark to answer. The main question being, how in the view of all that had happened during the last two days, could he know more about God than the new disciples? John put that very question to him. He was also told that every station in the world was now picking up their programme and this was going live worldwide. It was the biggest viewer numbers since the first Moon Landing.

During the add break, John asked Mike to take his place as he was starting to feel unwell, but gave Mark his promise he could have all the time he wanted. He suggested Mike might need a small sweetener to ask his questions and go along with him. During the break they also had to make a few changes to their signals so everyone would pick up their station. Mark smiled at Mike and handed him a cheque for a million pounds which Mike took and agreed to let Mark have his say, as he thought about it, he would be a fool not to. He knew Mary wasn't here yet and no matter what happened, Mark would get his word out just like Mary would the following day.

"Welcome back to our special programme, I am Mike Adams and taking over from John while he attends to our worldwide viewers. This programme is now going live throughout the world and we are all waiting with anticipation for Mark to answer the last question that was put him. So Mark if I may put the question to you once again, how can you know more about God than the new disciples?" Mike asked with a new vigour.

"I have studied the bible and seen that through world events, no God would allow His or Her children to call it something else. If God existed at all, He or She would ensure its children were well cared for.

"How on Earth could it sit back and its haunches, if it has any," he sniggered. "And watch deprivation, starvation, war and atrocities take place; killing not just the odd man here and there, but thousands upon thousands of soldiers, women and most of all, children. Children who have no say in the matter, they were too young to vote, but not too young to die. Children who died at the hands of faceless pilots who dropped terrifying bombs from the sky, mowing them down like blades of grass with machine gun fire from their planes of war. They killed people indiscriminately, without a thought or care of what happens next, the pain and suffering they go through, not just for the time of war, but if maimed, for the rest of their lives, where a society doesn't even look after the very people it helped ruin their lives, so say for their country. At times, a country owes it to their people; yet this country is one of the worst in the world."

At that moment, scenes of the Second World War were shown on the screen behind them and the viewers could see them with the people in a small box on the bottom right of the television. They saw bombs being dropped on England and Germany alike, people being blasted to pieces, others being carried half alive, with an arm or leg or both missing from a demolished building. The film showed people running on fire from a blazing building, then close ups of their faces and hands burned to a crisp, screaming in agony and pain. People were shown with their stomach opened to the elements, dying before those watching in their homes. In the background were adverts of Britain Needs You, and underneath, people dying covered in blood.

Many people who lived in those times did not even vote for the party who took them to war and expected them to go and fight for their country and more than likely die in the process. The film moved on showing queues of disabled people at DSS offices and Job Centres throughout the British Isles. They were hoping to get financial help but were turned down; and those people sat behind their desks suggested they get a job and build themselves up to fight in another war and hopefully die and lessen the debt to the economy. Those people had fought to allow the people who sat before them to live in peace and even be born into a Free World, were now forgotten, a race to be swept under the carpet.

Finally the bomb to end all bombs: the atom bomb that fell on Hiroshima and the final downfall of Japan by America. Scenes showed fire and incredible heat racing through the streets ahead of the firestorm that was hot on its heels. Homes, as if made of very thin rice paper were gone in less than a second. They saw in slow motion, how people in the line of the heatwave had their skin and hair burned from their bodies whilst still alive and trying to run away from the hell on their heels. As they took in a breath of boiling hot air to scream, the heat incinerated their lungs from the inside out. Their skeleton still running until it too was consumed not by fire, but the searing heatwave running far ahead of the fireball.

The fire consumed houses, streets, roads, factories and even metals of all types melted in seconds when it came into contact with the fireball. It made no difference to the hurricane racing over the land; anything in its way was consumed and thrown up into the growing mushroom of heat and fire, travelling miles into the air. The pit in the middle where the bomb detonated was getting deeper and deeper as the radioactive elements burned the ground and altered its molecular structure, the pit of destruction and doom.

After the fire came the wind, the wind put out all the fires which the bomb had started; lifting man, machines and buildings, what was left of them, high into the air, helping to enlarge the ever raging mushroom cloud of death and destruction and then smashing the items down upon the land with such devastation that everything was flattened.

Finally, worst to come than the destruction of the city and the fire of hell and searing heat, worse than the raging wind with its power of destruction, was the radiation. Radiation spreading invisibly on the winds, contaminating the land, making everything it touched inedible and poisonous, not just for a few days, but decades. For those that didn't die in the initial high level poisoning, the radiation gave life a painful death and when they had children, their DNA passed on some of the radiation poisoning to their children. The world watched, seeing what they had unleashed upon their fellow human beings and relished the fact that it brought an end to the second world war, but at what cost to human life, humans who were in the wrong place at the wrong time. Man should have learned by their mistakes, but they built bigger and better bombs, ready to strike at a minutes' notice.

They said it saved lives, thousands of lives, but how many people died in the holocaust of Hiroshima? Soldiers' lives were saved, so too the politician's, but it was the women and children who died in their place; it was them who paid the ultimate price for their right to live.

With the final pictures being shown of mutilated children, scenes which had never been seen before on film as if it had just been taken, no fading or alteration of colour, it was like seeing it in HD; the people in the audience were crying, even Mike and John were crying, they couldn't stop themselves.

People called the station in tears, saying how sorry they were and begging forgiveness, but they didn't know who from. Even politicians called to say how sorry they were and admitted they had to do more for the people now in homes and injured war veterans, no matter what war they fought in, they would no longer be forgotten, the Prime Minister said so himself in a live video call over the air. The world's population appeared to be sorry and had tuned into this one station, which was now showing to over a billion people as they were being woken in the middle of their nights' sleep to turn on their TV and see what was being said; Mark had the whole world in the palm of his hands and loved it.

"With the dawn of the atomic age and the destruction of Hiroshima, I ask you, would a God really allow this to happen in Her or His world? If you were God, would you allow this to happen to your subjects; surely for the price of peace, you would have stepped in and did the job yourself?

"God would like us to believe He, or She has given us everything. If that were so, She would have given it to everyone, not just to the chosen few rich people among us today. Not just to the rich countries, allowing poorer countries' people to starve and die on land which will not grow food or support life. God would have made everyone equal; She would not allow wars, starvation, rich or poor. Hot countries, cold countries, lands of plenty and deserts which grow nothing but misery and despair; I tell you all now. . ." He looked seriously at the cameras surrounding him and spoke with a voice that penetrated to everyone's soul.

"There is no such person or entity as God. There could not be and never has been an entity of any form that could have created the heavens and Earth and above all; Man. It all started as the scientists have proven with the Big Bang. From there, life existed over the millions of years which it took to grow from the simplest amoeba to the being we now call man. We will grow from what we are into something more beautiful, it will take time and I assure you the female so called daughter of God; who is to come to us in eight hours time is no more than the daughter of God than I am the man in the Moon." He paused a long time looking about him at the silent people who were now listening intently to every word he said.

"She is no more than a woman who has perfected the use of her mind. The church, lacking in numbers and support has come up with this brainwave. It's a hoax to beat all hoaxes, to fool not just the people of Great Britain, but the entire world," he said aloud with a firm and loud voice, sending the words deep into everyone's mind so they would not forget what he had said.

"Mark, if I may ask you, what of the church? The fantastic scenes that have taken place since God spread Her hand over the planet?" Mike asked calmly.

"It's a mere trick, an illusion; I will give you an example Mike; tell me what you would most to see here and now," he asked with a smile. Mark had searched deep inside Mike's mind long before he had a chance to think of the question put to him.

"My daughter Susan; just before she died in hospital at the age of four; that was over ten years ago," he sighed, looking very upset at the thought of his beloved child.

The studio instantly changed to the hospital room where Susan was five minutes from her death. Mike's wife, Joan was there with the doctor and nurses trying their hardest to save Susan's life after she had been hit by a speeding car which had went out of control and failed to stop.

"This . . . this," Mike paused.

"This is your daughter Susan, go to her side, by the way the car was a blue Ford Escort, registration number XP 06 YAT, remember it for all time, your daughter's killer is Mr David Peterson of 36 Westmares Drive, Lincoln; he was 22 then and drunk at the wheel of his car. Go to your daughter; say your farewells, something you did not get a chance to do at the time of her death," Mark said looking at Susan in the hospital bed.

"Daddy's here my love," he whispered into her ear. Tears streamed down his face into the heavy bloodstained pillow. He bent forward, kissed her blood stained lips and gently, softly, stroked her brow. She was real; she was there in the flesh and blood. Somehow; he neither knew nor cared how this had been done, he was just grateful for the chance to see her and would only wish the surgeon could have saved her life. By the colour of her skin alone, he knew she was dying and there was absolutely nothing anyone could do to save her.

Susan's eyes opened for a brief second, she recognised the man before her as her father and was glad he was with her. She forced her pretty face to smile despite the deep cuts and bruises where her head collided with the steel lamppost. Her right arm and leg were also badly crushed and torn open to the bone as the tyres accelerated over them as the driver tried to escape the accident. He was not even aware he had hit the girl and his front bumper was crushing her against the lamppost as he forced the car forward, the front left wheel was spinning over her leg; tearing her skin and muscle from the bone as the front wheels rotated going nowhere.

"Daddy," she whispered, "I love you . . . hold me . . . please," she said quietly.

Mike gently held his daughter in his arms, hugging her close to him, despite the cry of pain she squealed from her blood filled mouth. His tears were no longer under his control; his emotions were overflowing, uncontrollable grief for his beloved daughter. Another doctor entered the room and looked at him in silence; he took the girl's pulse and waved his head in despair at the nurses and her mother.

She was gone, her internal injuries were far too dramatic to try and resuscitate her. It would have been a complete waste of time, the girl's heart had ruptured and her brain was badly damaged, how she had managed to utter a word he had no idea.

The audience looked on in total silence; they had no idea if this was a trick of the cameras or the real thing. The way things were going, the way Mike had reacted, far better than most top actors could have done; everyone knew deep down, this was at least ten years ago.

The police had hundreds of calls saying they knew David Peterson named on television. Many people recalled he had looked shaken and the car was badly damaged for over a week before it was repaired like new. Within ten minutes they had the new address for him and a huge crowd was forming outside his house and people were getting ready to lynch him. He was so frightened he had to lock himself into the house with his family and wait for the police to arrive and arrest him. He called them ten times, begging them to hurry as windows were smashed around him and he was heard screaming at the crowd outside to let his wife and two small children leave in safety.

By the time the police arrived and arrested the man, they had to take the woman and two children to the police station for their own safety as the crowd was so threatening. As soon as they were taken away, the house was broken into and everything inside destroyed. It turned out David Peterson's wife knew nothing about his former past as she had married him five years ago after going through a divorce and had her two children with her, for the time being she had to stay with her parents as she could not return to her house until it was repaired.

Mike was given all the time he needed to be with his daughter and John joined him, all those years ago, after Susan had died. When the scene gradually faded, his daughter faded with it into his memory. Mike was once again in the studio and broke down in tears, not of agony and anger as before, but of relief that he had been given the chance, no matter how it had happened, to say goodbye to his daughter.

"That was some illusion Mark," John said standing by Mike's side. "Would you care to explain how you accomplished it?" he asked.

"It was no illusion, I saw deep inside Mike's mind what he wished desperately to see and we saw it, as it happened in real life, except at the time of the accident, he was on location in Scotland, working for this station as it happened. You didn't tell him of the accident until he completed his evening's work and prepared the interviews for the following day. You held up his return to his wife and child as long as you could, just to get him to do his work which you needed to be completed.

"Do you know, if he would have been allowed to use your helicopter, he could have caught a connecting flight and been there with his wife and at daughter's side when she died?" Mark explained.

"I didn't think it would matter, I was told she was almost dead and would not make it through the night. I needed Mike to. . ."

"You needed him more than he needed his wife and daughter or they needed him. You took one man's life for your own greed, but you never told him the truth, that is, until now. Perhaps he will feel differently about you and the station now? I have given him the opportunity to see his daughter when she died, you didn't give him the chance to try and be there." Mark said with a smile.

At that moment, Mike turned and hit out with all his might at John's face who then flew across the room and hit the floor hard.

"I owe you that much," he stated and went to walk off the stage.

"Stop Mike," Mark commanded. "I have given you those last few precious moments you desired, you will be the one to show the world that God does not exist and this . . . this woman called Mary is no more than an Earth woman with many mind talents."

Mike stopped in his tracks as John stood up rubbing his bruised chin, this could be the scoop of the century and he wanted to get in on it. If Mike was destined to be the number one reporter, he wanted to be the number one producer and nothing was going to stop him. He walked up to Mike, looked him directly in the eye and spoke, before the billions of people watching around the world.

"Mike, this won't change how you feel, but at the time I was told she had less than an hour to live, she wouldn't have made it through the night, that was certain. You were on air, it was a hard decision for me to make, and it was the hardest decision of my life which I deeply regret I got wrong. I had no back up and we were just starting out, if we messed it up, we would have lost our place in the ratings and the show would have been cancelled for good.

"I made what I thought was the right decision on the information given to me by the doctors at the time. I immediately asked for a replacement for you but Don Chapman couldn't get to us until the following evening; he wouldn't have had time to make his notes. I made myself scarce on purpose and gave strict instructions you were not to be disturbed or have any communication from the outside until I had spoken with you. By then, I had managed to make all your arrangements to get you on the next possible flight to Bristol. I've had to live with that decision for all these years and it hasn't been easy. Please stay, you have a great name for yourself and a job which I would trust to no other person except you. Stay. . . do what is yours to do," he held out his hand, hoping Mike would accept his apology, but deep down knowing he wouldn't.

"Okay, but when this is over, we're through I can't work with a producer who isn't straight with me or anyone else for that matter." He refused to shake his hand and took his seat and ushered John from the stage shaking his head in disgust.

Upstairs, where the directors watched from their soundproof booths they discussed among themselves who would be the man to go. They had ideas of their own, and knew Mike was their best man, the money puller. If the station was to survive, a new producer would be found who Mike could work with and have complete faith in.

They knew they would have to convince Mike they had nothing to do with John's decision, they would say they ordered him to be told as soon as possible what had happened to his daughter and get him to the hospital as fast as they could. Mike was delayed purely because of John; it was John who would carry the can for this miss hap.

In his mind John knew he had won the first battle, now he would have to work on Mike to keep him and his station running. He looked up to the director's room seeing a look of dismay on their faces and began to sweat profusely.

The interview continued for a further thirty minutes, running well into another top programme, but the interview was so good, the directors decided to over-run and let the world's population judge for themselves if Mark was telling the truth or a lie

CHAPTER 10

While Mark gave his interview, First Lady Rocc discussed matters with her disciples before sending them to their various churches to prepare for the second coming.

"The Anti-Christ is among us, go with caution, beware of whom you discuss matters and most of all, take care of yourselves. Until Mary is with us we are at Mark's mercy, despite the powers we have been given by God, they will not stand up to Mark's powers as his come direct from his Father, Lucifer. From this moment forward, we will no longer call him the Devil or other known names, but give him the respect He deserves and use His name being Lucifer and let us hope this may get us some leniency from Mark.

"While I have been talking to you, Mark is putting God down before the world. I have tried to teleport into the studio to put our side over but He is much stronger and has stopped me from doing anything to interrupt his speech. He has erected a barrier, a force field of a sort to keep us out, and that also includes the police and other authorities who have tried to gain access to the building. They have even cut power to the building but it is still lit and he is speaking to every person in the world, saying Mary is no more than an Earth woman with special mental powers.

"Of course we know different, we have been waiting for her to come and she will be with us in nine short hours. Until then we must keep our faith even though it is being changed before you.

"I know this is very out of the ordinary, but I would like to ordain the three ladies in my group and Cannon Mathews to Bishop. I have been asked by God to do this; does anyone object to this proposal?" First Lady Rocc asked setting the congregation to chatter.

The abbey was soon silent and not one hand was raised during the ten minute recess.

"Archbishop Langley, I would like you to conduct the service immediately if you would please."

"I would be most honoured," he replied. "I will require five minutes to set the service up. I take it we will dispense with all the formalities and get straight down to the service?" he asked.

"If you would please be so kind," she replied smiling.

After the ceremony, Mrs. Peters, Mrs. Meadow Miss King, Cannon Mathews and as an added extra, Dean Huntley were all ordained as bishops into the new church.

"Now the main work has been done, there is also a very special post for Bishop Mathews," First Lady Rocc stated and immediately, the bubbling happy small chatter silenced in the abbey.

"As Bishop Mathews has done so much research and understands what is happening abroad, he will be put in charge of overseas research and establishing the church in Israel for the next year." She paused as voices grew louder within the abbey until she could no longer make herself heard.

"Brethren please!" she shouted into each member's mind. She hit them with a single thought, which brought the entire congregation to sudden silence.

"This post seems to have brought about a great deal of talk. I assure you that this post was not of my choosing, but that of God's. It is God Herself who has ordered me to make this post for Bishop Mathews. Do not forget, he is by God's choice a disciple; disciples spread the meaning of God's name and what she will do. This time we will be making vast improvements for the time Mary is among us. Mary is not going to stay forever on our planet, but Her memory will, so will that of the True God, and she will not be crucified by the people of this world or by Mark."

A large cheer erupted throughout the abbey."You must understand; God wishes all races to believe in Her, that this female God is the creator of the universe and not the false God of the male species. There will no doubt be new laws the church will administrate, I am sure in this day, God will see fit to punish those Herself who dare doubt Her word."

The abbey burst into voice again until Archbishop Langley silenced them.

"In less than two years woman and man will be ready to leave this planet and travel far into space and search for other life forms; to take the name of God out to the stars and other universes which you have only dreamed of.

"There will be a leader for this mission, a leader of the church who will carry God's word into outer space. This person will be of great courage that has showed understanding and pity to others in the past. He has been searching all his life to find out the truth," Archbishop Langley stated.

"Who will it be?" voices asked throughout the rows of the pews. Everyone was eager to know, all happy that man and woman would eventually go into space as Christians and spread the word of God.

"The person, who will be in charge of the entire mission, will be our new First Lady of the Church, Mrs. Rocc."

A rapturous applause and standing ovation followed for First Lady Rocc, as everyone was extremely pleased for her. At the same time they were slightly upset that a man would not do the job but in the light of the new era, the fact that God is a woman, they were not surprised. The fact now that God is female made it obvious a woman should be in charge.

"At her side will be a man of equal stature; he will be the most ruthless man in our church today. A man who despite other people's views and principles, persisted in his task failing no one, and stopping for nothing. I expect most of you know who I am talking about." First Lady Rocc paused for a brief moment looking about the abbey, as if searching for a man among the rows of pews to see where he was. He was actually standing behind her, deep in thought about his mission to Israel.

"This man," she paused again to get everyone's attention, "Is Bishop Mathews." She moved to the side and turned to face the man she had just named who had not heard her. The man of the moment was still deep in thought that he didn't know anything about his new position until First Lady Rocc spoke to him using telepathy.

He understood what was happening instantly upon having made contact with First Lady Rocc and he was fully informed of his second job into outer space.

He looked to Mrs. Rocc and smiled, together they stepped forward to the loud applause of the congregation. The ministers were thrilled; thrilled it was not them that were being sent into deep space for this very hazardous job. It was one thing to go to another country, but in space. . . man had barely gone to the Moon.

There was a permanent base on the Moon now, despite the fact it was the army in charge and it had only become known to the public over the last two days. The cost had been colossal and MARS ONE had managed to get their first twelve astronauts to Mars and despite NASA's fears the mission would fail, the men and women were still alive and awaiting the arrival of the next twenty crew and more supplies from Earth.

Many members of the church had agreed with these people saying it was wrong for man to leave Earth, yet now, here was First Lady Rocc telling them they were bypassing the Moon and Mars and in two years, would be travelling to the stars on craft that could go faster than the speed of light. God had made the stars for women and man to see close up and visit; and the people of Earth would now see their planet as a country in the vast world of planets, with oceans of space separating them.

It was the most radical change since the understanding of the bible and the female God. Man was learning so much and so many of his laws would have to change, not in the long term, but overnight; new governors and new systems, bringing women to the position in society they belonged to and was theirs by right.

As the blessings took place of the new positions, all at the same time, a choir, much sweeter than anyone had ever heard before, much softer to the ear, all in perfect harmony, sang in the church.

Archbishop Langley stood back as a brilliant white light surrounded the congregation. The choir grew louder and the music which accompanied them; pipe organs, harps and flutes filled the abbey with their throng of delightful praising music. The congregation stood instantly, and looked in disbelief at the apparition before them. They had heard the stories of what happened at Radstock Church, now it was happening again before their very eyes. There was no doubt this could only be the work of God. In Her own house, She was entitled to touch whom She so desired.

Everyone knew they were being blessed in a very special way; baptised for the amount of time they were bathed in the soft white light. Many held their breath, afraid they would miss something should they breathe or blink. As gently as the light came down, it lifted away and passed into God's own world; it was only when the light receded people saw the entire group of disciples and had been bathed in the light. Then it dawned on everyone in the congregation that they too had been baptised by God and touched by the light.

As each member of the congregation looked down, a new bible was in his or her hands. It had a pure white hard cover inlaid with gold print. The pages were the same gleaming white with jet black ink and each chapter heading was inlaid in gold. The page edges were in gold and the bible was thick, it should have been heavy for its thickness, but it was light, very light indeed.

As each person turned the pages of their new bible they were all stunned at how fast they could read, not one person, but everyone in the abbey could read at an astonishing speed recalling every word, every story, parable or speech. There was one other major point about the bible, there were many coloured members in the church and in the congregation representing other races, when they read the new bible; it was in their own language. When they pointed it out to their neighbour, they shook their head saying it was definitely in English, or whatever their native tongue was.

They could speak every language on the planet and somewhere inside, something told them they could do much more. It was a long quiet moment before First Lady Rocc spoke again; it wasn't that she wanted to give everyone the excellent news, but as people could plainly see, she was being spoken to by God. Finally, she opened her eyes and praised God on high.

"We will pray together, thanking God for all the wondrous gifts She has given us this evening."

She led the prayer, short and to the point then looked up, glanced at each individual person and at last spoke.

"God has touched every one of us here this evening; you all have the power to read with phenomenal speed recalling every word you read. There is something else; all of us here have been gifted with extended life, that is, life beyond our normal expectancy."

"How long?" Reverend Peter Holland sat in the second aisle asked.

"At least to the age of two hundred years, those chosen for specific tasks will live for much longer; it is the price we shall have to pay for being God's disciples this time round. In Jesus' time, the disciples did not live long enough to spread the truth about God and somehow, in the evolution period, man lost the truth, God has decided this will not happen again. Her disciples will live until the true word of God is spread and remembered throughout time with no more lies being told about Her.

"You are all Her disciples; chosen this night by God, you must be ready to go where God sees fit to send you, knowing you will be protected by Her."

There was much talk at this point as people discovered they were now disciples of God. They were also clothed in bright white suits and no matter how they tried to put dust on them; it didn't make a solitary mark on God's clothes

"I thought there would just be twelve disciples as in the days of Jesus," said Dean Green, who had just stood up to take a close look at his new clothes.

"There will be many more disciples this time, more women than men. Some people in a few years to come will not stay on Earth but go out to the planets and stars, and even distant galaxies and new universes as our ships get bigger and faster and we start to fold space, and I don't know where that came as I am no astronomer or astrophysicist, or at least until a few minutes ago. You will all help spread the name of God far, very far and wide."

"How will we know when our job is completed?" a minister asked from the rear of the abbey. His voice though quiet, was clearly heard by everyone in the abbey.

"When God has finished with us and welcomes us to Her Kingdom," First Lady Rocc replied.

"Then the world will not end tomorrow?" a new female disciple asked halfway along the abbey, again her quiet voice was heard by all.

"No my sister, the end of this world will meet its natural disaster when our star dies in about half a billion years from now. Only at that point will our planet die, and when that time comes, we will be far away from here and living with our friends in God's name in peace on planets throughout the Cosmos.

"This is a confusing time for many of us, some of us are married like myself and we have been chosen. What is to become of our wives' husbands and children? I have three children to care for," Disciple Christine Mallard asked, not just for herself, but for everyone in the same position as herself.

"God will answer all your questions when the time comes, for the moment; let us concentrate on our immediate problems which lay before us. I promise you that all your families will be well cared for, any illnesses they have will be cured and they will not worry for you or themselves, as they too will be filled with the spirit of God. Mark the Anti-Christ is among us, here on Earth before Mary has had a chance to step foot on our world. He will do everything possible and his power is very strong, to prevent tomorrow's events taking place.

"It is very important that every disciple gathers at Radstock Church as soon as this meeting is over. Your families are safe and you will feel their warmth and guidance beside you. You will also be able to speak with your families using telepathy to give you further reassurance," First Lady Rocc told everyone.

We all noticed how the light remained on you for a considerable time. Can you explain what extra gifts God has given you?" Reverend Knight asked from the back of the church.

"The main gift for everyone here is to travel to distant worlds, to understand other creatures and be able to speak to them in their own language which we can all now do. This can only mean one thing; the galaxy is populated beyond our wildest dreams: we are not alone in God's Kingdom."

Silence filled the abbey as the last of the questions were answered. It was now almost midnight and many people had to travel vast distances to get home and then get to Radstock Church; time was now against them as ministers and disciples alike looked to their watches, they wondered if they would return to their vicarage in time to welcome Mary to Earth.

"You will discover when you get to your churches, all your bibles have been changed to the new bible, in fact, all bibles throughout the world are now of this nature and speaking of the true God. It is time for you all to leave and go your separate ways. Trust in God and have faith, go now, spread the word of God to your congregations. The one hundred special disciples, and they will know who they are, please remain behind for a few minutes."

The group watched the congregation leave their seats and silently leave the abbey. As each person stepped out of the abbey, they found themselves stepping into their own vicarage and the transport which they used to get to the abbey was waiting there for them. As they turned around, there behind them were the abbey doors, closing silently in the distance. Now each minister understood that only God could have accomplished this miracle. There was no doubt in their mind that God was with them; everything First Lady Rocc had said was true, they felt exhilarated and praised God on high as they stepped quickly forward to their homes to prepare themselves for the coming of Mary in just a few short hours.

"Tomorrow there will be thousands of questions for us to answer and we must be there to answer them," First Lady Rocc said with a sigh when the abbey was partially empty of clergy.

"Do you think they believe in the new God, I mean the change of sex, so to speak?" Archbishop Langley asked with a cheeky smile.

"I certainly hope so and by the time they get home, which they have all done as I speak, I'm sure they'll believe."

"So First Lady Rocc, what are we to do now?" Bishop Meadow asked.

"Return to Radstock Church, I have a feeling that if Mark can stop us getting into the studio, then he will do his best to stop us, or least try to detain us from returning to the church."

"What of us disciples?" Disciple Catherine asked.

"You have all been given destinations by God to attend for tomorrow's event. They are critical to the event taking place, you will still see everything in your mind as it happens; as if you were there in body and it will not deter you from your task."

"But, First Lady Rocc, I am to go to a prison where there are ruthless murderers," Disciple Marian said bashfully.

"I too," said another four disciples.

"I am to attend a hospital," said Disciple Maureen.

"I know where you all are to go," First Lady Rocc interrupted, before everyone told her of their destination. "You are being sent to places to cleanse people of their sins and illnesses. It is God's wish that no person on Earth should be incarcerated; you will be the disciples who will release these people from their nightmare bonds. You will clear their minds, ease their thoughts and turn them into good Christians. Every prison Governor is expecting you; they wish to see if you can accomplish your task and see the changes in their prisoners. This will happen at the same time as Mary sets foot upon the Earth. When you have completed your job in each prison, you will move to the next and God will open the prison gates and allow the people to walk free and change the building into something which God wants it to be; so please move quickly in your tasks. Do not linger too long; although you will be asked to, you have many people to set free tomorrow.

"Those of you who are going to hospitals you will do the same thing; heal those who are ill and about to die. When the healing is completed, God will cleanse and bless each hospital you attend and give the doctors their new gifts to help them heal people. Now my disciples, go in peace and with speed, you will find it easy to move about like you need to." She smiled as one by one the disciples left their seats and arrived a moment later at their destinations.

"It is time for us to depart as well," Archbishop Langley said urging the others to form a tight circle. They knew by travelling in large numbers there was far less chance anything would go wrong, especially with Mark's intervention. When the last people disappeared from the abbey, it was transformed by God into Her new house.

The verger, stood at rear of the abbey, amazed at what he saw, he knew the group had powers, but had somehow never believed the story he had heard with own ears. Never before, had anyone, including himself seen people travel in this way. Now he was seeing it with his own eyes, he would doubt their word no longer. Lifting his head he prayed to God for Her divine help in their future quest as God refurbished the abbey around him.

To the groups delight they stood outside Radstock Church; all around them were camera teams and vans with large satellite dishes on the roofs and even larger dishes set up around them on the ground with cables snaking everywhere, ready to send the pictures around the world.

Some of the crews managed to capture the group's appearance which immediately started loud chatter among the commentators. One man, Tony O'Reilly a commentator from Sky 5 News ran forward, hoping to be the first to get an interview with First Lady Rocc.

He checked behind him that everything was working, ready for him to start the interview, if he could. Knowing what he was about to say he approached First Lady Rocc.

"Good morning First Lady Rocc, do you have any reply to the allegations Mark has made about yourself, God and the coming of Mary?"

The group were stopped in their tracks, stunned in silence for a moment. "Allegations?" First Lady Rocc asked clearly. I have not heard any allegations," she replied.

"Surely you heard what he was saying on television last night about you being nothing more than a normal woman and the new Christ Mary, being nothing more than a mortal; a sort of super woman with special gifts like telepathy and telekinesis. He is saying you have mental powers developed here on Earth and you are now being used by the church to boost its numbers."

"We are not aware of any such allegations," Archbishop Langley confirmed. "Tell us, when did these allegations take place?"

"Last night between 10.00 pm and 01.00 this morning." He looked at his watch seeing it was now 3.15. "It's just two and three quarter hours to go before your Mary sets foot upon the planet Earth. I ask you seriously, because Mark has put a lot of doubt into people; is she just a woman with abnormal powers?" Tony asked showing concern and worried this story was nothing more than a story, to help bring people into the church and swell their numbers. He believed in God, and wished it to be true.

As if to answer his question, the sky was lit with brilliant electric blue lightning, the thunder which followed was almost deafening.

"I think that answers your question," First Lady Rocc stated. "She is not and will not be of this planet. She is the daughter of God; during the time you mention, this group of ministers were conducting a service in Westminster Abbey, talking to over 800 brethren of the church. We knew Mark was up to something, but as he had refused me entry to the studio, I did not witness the allegations. However, if you will allow me to enter your mind, I will bring myself up to date and at the same time everyone who was with me at Westminster Abbey will also know what he has said."

"If you think you can do it, personally, like many others now I have my doubts," he said with a smirk on his face and chuckled to his crew, making a screwball sign with his fingers.

At that moment he felt something inside his mind, it was a feeling of not being alone any longer and having his mind opened up for all to see. He knew instantly by some means or other, First Lady Rocc was somewhere inside his head. Try as he did, he could not stop her and he was frightened of what she would see.

When she left him, a brief moment later, she smiled. "Well, I always wondered what was inside a commentator's head, not much it seems except where you are going to have your next drink and a meal," she said making him wonder what else she had seen. "We are now up to date with what Mark has said and I would like to make a statement."

Tony looked at her, got his camera and soundman saying everything was fine, as other reporters were also trying to hear First Lady Rocc's comments.

"Mark is the Anti-Christ, it is foretold in the bible that he will come just before the second Christ; it is Lucifer's way of testing the true God. To put Her down; now we understand the truth and Mary will tell us in her own words what she wishes us to do. What Mark has said is lies, I am no more a super woman than the next woman, but I have been blessed and given powers by God to help preach with, to pass on the true word of God to others and to travel quickly around the world."

Tony was still recovering from his intervention with First Lady Rocc; he never believed it possible; now as he saw the light, he actually believed what was being said. Without a shadow of doubt, he was on God's side and he knew he would help prove beyond all doubt, no matter what his producer said, or how he wished the programme to go, he would prove there is a God who is female.

CHAPTER 11

Sarah walked over to Tony and made him more acceptable to the cameras, wiping the sweat from his brow and brushing his hair. She got more colour back into his face which had suddenly been drained by First Lady Rocc's look deep inside his head; a part of him he assumed nobody could get to.

"You look terrible, are you sure you're alright?" Sarah asked in a caring manner. She had looked after many artists for the station and Tony was one of her favourites.

"Yes I'm fine Sarah," he replied, recovering from his ordeal. Even though, he still seemed a very long way off, deep in thought, as if planning his next move.

Tony raised his arm gently pushing Sarah away looking deep into the mirror before him which had been quickly placed on a table by the makeup artists. There was something else inside him now, something that wasn't there before, although he couldn't put his finger on it.

"That's better," Sarah said with her usual smile, still fussing with his hair. "You look brilliant, handsome, in a strange sort of way. Have you done something to yourself?" She could see it now, Tony looked radiant, he seemed to glow.

"That's it!" exclaimed Tony, looking ever deeper into the mirror. "I am!" he continued in an almost silent voice.

"Am what?" Sarah asked bewildered.

"Never mind Sarah, that's all for now, tell Adrian; who calls himself a producer, I'll be ready in one minute." He discarded Sarah with a smile of warmth and affection, his hand sliding off her shoulder as he looked again into the mirror, moving his face forward so he could get a better look at himself.

There appeared to be a glow emerging from inside his face, reaching out, filling his body with a new form of hope and understanding. When he sat back in the chair, he starred at the image before him.

There was an aura about him which melted into the night air; despite the makeup, he looked different, from inside out. As his mind surged through his body, he realised he could only have been touched by a messenger from God.

"There really is, I am the living proof that miracles can and do happen, even in this day and age," he said silently to himself.

"Tony, time to go on again, now, I want you to start on a new angle," Adrian explained. Then he stopped and looked at Tony. "You look different, sort of . . . I don't know how to explain it." He searched his mind, looking for the correct word among the millions he had used over the years. "Radiant!" he stated suddenly, realising then only women looked radiant, but there was something in Tony which shone, he had changed. His face had a warm glow, a soothing glow which was pleasing to the eye, he was indeed; radiant.

"Thank you," Tony replied, accepting the compliment and stood to face Adrian.

Then Adrian saw something totally different in Tony; his nine o'clock shadow had gone. He knew there had been no time for him to shave and makeup was never that good, not out here in the sticks. It took hours to transform a shaven face into the smooth soft features of a woman's or child's skin. Tony's face was a like a child's, fresh, re-born, soft, like new, and it glowed, with a glow he found hard to describe. Except, except no glow like this could ever have been made on Earth. It was so pleasing to see, warm and welcoming, like you knew there was nothing to fear from this person, he was not about to lie to you; this man didn't know the meaning of lies.

When Tony spoke, he found himself listening intently, hanging on to every word, despite the fact he wanted to convey his thoughts on how the remainder of the early programme would go now the Anti-Christ was in the picture. Adrian wished to put these women down and prove them as frauds; just like other people he had done to in the past. Now something had changed his mind, Tony had changed and he didn't or couldn't say a word until Tony had finished speaking.

"I'm so positive that what has been said by First Lady Rocc is correct, I am willing to stake my reputation on it. I'm sure you would like to see me wrong, but now I feel different."

Then Tony realised Adrian had not interrupted him like he usually did, so he continued to speak, getting out his side of the story before Adrian superimposed his side on him and had his way.

"Since she entered my mind, a shattering experience I may add, I feel reborn, like new. It's as if. . . . I'm another person; you know the feeling when you go to church? Not the old church with its congregations of people in suits and pretty dresses, I mean the New Anglican Church, where people can be re-born unto Christ. I feel. . . exhilarated, glad to be alive. I really do believe in God, like I've never done before, despite my being a Christian Baptised into the church.

"When she came into me, she didn't just read my mind, she wiped my mind clear of all impure thoughts. I feel incapable of being able to tell a lie, even a white lie to please my wife, like, she looks lovely when she really doesn't. I love the dress she's wearing when I really hate it and wish she hadn't bought it on my credit card; do you know what I mean?"

"Yes," Adrian muttered, almost in silence as he just stared at the person standing before him.

"I know what you're thinking about, this Mark, the Anti-Christ chap, but really, he's a sour note in our time. He came to manipulate people to say what he wills; to make us believe what he wants us to believe, that bit with Mike on the television last night was a put up job. Oh yes it all happened, but after all said and done, if he was truly able to work miracles, he would have brought Susan back from the dead. Now that would have been a miracle and God doesn't need to work in miracles like that to prove She's a God, to prove She's alive in our thoughts, spirit and hearts.

"I know this is going to sound really stupid and you are the producer, but I would like to handle it my way. To give these women all the help they need; fighting the Anti-Christ is not going to be an easy job and it's up to every one of us to lend a hand and believe in God Almighty. Are you with me or do I do it my way by myself?"

Tony waited patiently, knowing Adrian was being pushed from people on high who ran the station and only wanted to get as many viewers as possible. This one event would have the entire world watching a screen of some sort. There were so many channels here, the airwaves would be congested with UHF, VHF and satellite signals. It would be worse than spaghetti junction in the worst rush hour ever.

"Ok Tony, do it your way, something tells me you're right. Something I call instinct tells me God is going to be on our side and I still can't understand why Mark hasn't turned up here to try and stop this miracle from taking place. If he is all that powerful and these women are just women on the Earth, he should win hands down," he said holding his hands open aimlessly.

"Thanks Adrian, I won't let you down, you'll see," Tony said with a wide smile which lightened the night.

"You'd better know before you start," he paused for a long moment. "Mark has convinced most of the other producers and presenters to boycott this event and put those here down as much as possible."

"I have a feeling that when the time for Mary to arrive comes, the whole world will see it no matter what Mark says. I'm sure God is not going to allow Her daughter to be criticised by the press as soon as she steps foot on the planet to do Her work," Tony replied honestly. He smiled again, turned and walked away to have word with his camera and sound crew.

The group of disciples were now standing before the church doors talking quietly between themselves. Occasionally they turned to see a bustle of cameramen moving to a better position; none had any idea of how Mary would arrive, at that point. During the early evening, microphones had been placed in a huge arc over the church doors where they could pick up all the sounds which would come from that location.

To be fair to all, and to stop other sound men getting too close and ruining a perfectly good shot, the microphones were connected to a huge bank of outlets for all the soundmen to plug into. Even some newspaper press were allowed additional speakers in the streets, ensuring everyone could hear clearly what was being said.

The local council had arranged for giant television screens to be erected in parks so people wishing to get a first hand view could see. These were in addition to the screens God had supplied much earlier around the world. At the moment all the screens were blank; no one knew for sure if they would be turned on at Mary's arrival or if it would be later in the proceedings; so, just to make double sure, the extra screens were erected, despite the cost.

Everyone was checking their watches as the time came to 05.45. Cameras rolled for a test run on the group standing by the doors of the church and microphones were tested for the last time before the big moment began. What followed, no one expected.

"I am sure Mary will come in a peaceful way," said one of the women disciples.

"Oh no," said another. "With the sound of trumpets from on high and a choir of a thousand angels," she continued.

"You're wrong; Jesus came into the stable to the sound of his own crying voice and the bays of animals in the cold December night. I'm sure Mary will follow the tradition," said another disciple, for no one could see who was talking at the time.

"Come on, we're in the twenty first century; I'm sure God will send Mary in a more up to date mode than that," Archbishop Langley was heard to say. "I bet you fifty pounds she will come dressed as stunning woman, high heels, black stockings and short skirt, to get the young people interested in her," he continued.

"I'll raise the stake to a hundred pounds that she wears flat shoes and a modest length skirt, she's not a girl of the red light district is she," First Lady Rocc added.

"I think she will arrive in a white limousine with angels escorting her and open the car door for her," disciple Huntley argued.

"They're arguing," said Tony of the commentators. "I don't believe it, but they're arguing over the way they assume Mary will enter our world. I know it has us all stumped, but these are the last people I would have expected to hear argue over this and even bet money on the way she will arrive."

"Yes," agreed Ryan, another commentator from BBC News. "I would have thought these good people had better things to discuss than the way Mary will be first seen."

Then the group of disciples, on hearing what the reporters were saying to the people at home, turned to face the crowd of cameras; lights flashed as still photographs were snapped one by one; the sound of arguing could still be heard, growing fiercer and fiercer. The group looked up to the microphones and shrugged their shoulders; their mouths not moving, not one person in the group was speaking, at least not by mouth, which was the only way the microphones could pick up sound.

"Will you look at that, they're not arguing after all, in fact ladies and gentlemen, not one of them is uttering a single word and they look just as confused as we do. Our technicians are trying to sort out the mystery at this very moment," Martin, one of the commentators for Yorkshire Television explained to his viewers.

A camera panned around to show the scene behind the technicians trying to see how this was happening. The arguing still continued with shouting bad abuse thrown in, just to make it more thrilling. Then it stopped, just as abruptly as it commenced. Archbishop Langley and First Lady Rocc searched the minds of all the people close at hand. Then the people on the other side of the church wall. On one of the stands, erected so people could have a better view, they suddenly met a mental block . . . it was so powerful; they knew it could be only one person.

"Please, if I may say a brief word," First Lady Rocc stated above the roar of the crowd. Silence followed a moment later as cameras and sound technicians concentrated on her alone.

"What you have just heard are the many voices of Mark the Anti-Christ, who is doing as much as possible to interrupt our service here. We, your church leaders, would never argue over such an issue as to how Mary will arrive. She will do as God so desires and God alone. It is not for us to take bets on, or make enemies over.

"Mark is just being disruptive and trying his best to put us, as leaders of the church and disciples of God, to shame. He will not continue this charade and has just this very moment left the scene."

People standing next to Mark found themselves with more elbow room; he had not jumped down behind them or over the church wall; he just vanished into thin air, the same way as he arrived, yet nobody in the entire two thousand strong crowd had noticed his unusual arrival.

Without looking at her watch, First Lady Rocc knew it was two minutes to six. Just two minutes to go before they saw Mary. No-one knew how she would appear, or what she would look like or wear or what she would say to the sisters and brothers of this sinful planet.

The crowd was now silent; technicians checked their equipment for the thousandth time and camera crews focused their cameras onto where they assumed and hoped Mary would arrive. A voice in their ear piece counted down the final minute. Nobody was sure if this whole event was no more than an elaborate hoax; however, in less than thirty seconds they were going to discover the truth. The church would either live; or die the most embarrassing death it could ever wish.

If nothing happened then God was bound to be washed up; nobody would believe in Him or Her ever again, despite all the miracles which had taken place over the previous forty eight hours.

Tony knew Mark was around somewhere doing his utmost to bring this occasion to a dismal end; and he knew he would do his best to wreck the preparations which had been two thousand years in the making.

He also knew if Mary didn't show and Mark was indeed the Anti-Christ, then the Earth would belong not to God, but to the Devil, Lucifer, or whatever they would have to call him and certainly bow down to His commands.

Dawn was starting to break as the sun started to rise in the eastern morning sky. It was too low in the sky at the present to see, but the clouds were growing orange and flaming red as the bright rays pushed the night sky away and brought the first light of day which everyone hoped would be the dawn of a new era.

As the sky glowed brighter, the clouds appeared to disappear from the night sky very fast, as if welcoming something from the beginning of this splendid day. Everyone looked up as the sky brightened much faster than normal. The sun rose higher in the early morning sky and everyone knew this was not a natural phenomenon. As birds started to sing in the vicarage trees, more birds added to the early morning throng as they sang gaily, singing like they had never sung before; singing their hearts out to God Almighty.

Although it was late autumn, the tress suddenly came into full bloom, ready to bear fruit and not a single leaf fell from the trees. As people looked from person to person, on their right or left, they were in such good health, beaming in colour. They were suddenly their correct weight for their height and their clothes fitted well and were all brand new; they would later discover that all the clothes they owned fitted them perfectly and again were like new.

The miracle was only just beginning, Mary was still to arrive; yet all around them they could feel God's presence in the air. The air was so clean and fresh; every breath they took was like they were taking air from millions of years ago before it was polluted with today's modern odours. The air was now unpolluted, the ozone layer which had been damaged was healed and the oxygen was back to when the world was new and only just created. The clouds were whiter than white, gone the dirty elements which contaminated them, blocking the sun with dirt and acid.

The cameras, previously trained on the great doors of the church, were now scanning the wondrous miracle unfolding before them which was definitely not Mark's work.

"I wonder if this is happening anywhere else in the world?" Tony asked Bradley, his cameraman.

"I certainly hope so; this is one planet that needs to be cleansed; I'm sure it must be happening all over the world at the same time," he replied confidently, scanning the clouds, trees and even getting a shot of Robins in a tree only a few feet away.

Then he looked up around him; the entire churchyard was filled with brightly coloured Robins, more than he had ever seen in his entire life. Then over a hundred pairs of brightly coloured love birds flew into the blooming trees, settled down and began to add their voices to the other birds singing to God. He nudged Tony to say something into the camera as he finished his swing, showing the viewers the marvellous picture of Robins and Love birds which had come to witness the coming of Mary.

"Well ladies and gentlemen; you don't see scenes like this very often; possibly in a bird sanctuary which this isn't. Not only that, there are at least three thousand noisy people gathered here. Robins, the tiny little bird which visits our gardens in the winter months; have come in their thousands to witness the second coming and are landing right on our equipment and some are eating the crumbs our engineers have left behind while they set up our cameras," Tony said confidentially.

Richard Hall, the director of the programme tried to cut him off as they too wished Mark's view to get through to the people and disgrace the women and prove the church was making it all up. No matter how the technicians tried to stop the programme going out back at the studio, the signals continued to flow out as they were coming in and Tony was allowed to continue without further hindrance.

"For those of you watching in doors, or still in bed, put your head out of the window, walk into your garden or on your veranda, get outside if only for a few minutes, you will not miss anything, because you will be able to see the clean skies with your own eyes, feel the clean air inside your lungs and smell the fresh air with your nose. You will hear the millions of birds singing their hearts out as they welcome this special day, this special dawn, for what is said to be the new era for all humankind. God has cleansed the atmosphere and it's happening all over the world. The sky is a new colour blue, a much brighter blue which really shows the pure whiteness of the clouds. It's as though God has put a vacuum cleaner in the atmosphere and sucked up all the dirt from the air.

"People here have changed too; they all look so well, clean and healthy. This is not mass hallucination or some magic show, it's really happening at this moment, go look out of your windows and see for yourself, then return to your TV and see the morning unfold before your eyes.

"If this isn't a miracle of God, then I don't know what Lucifer is up to; He has never done anything like this before and Mark, His son, is nowhere to be seen. It's now ten past six and many people have asked where Mary is, but the miracle started precisely at six o'clock. I suppose we will have to wait for the miracle to unfold. God has commenced Her cleanup operation, sterilising the wind, rain and elements before sending Her daughter here. Cleaning up what we have managed to dirty over the years through war, coal, fossil fuels and now nuclear fuels, sending acid rain, radioactivity and chemical gases throughout the world into the air and polluting it.

"Nobody can say we didn't make this planet dirty and I mean dirty with a capital 'D'. We destroyed the ozone layer so much, we didn't care when we actually put our own lives at risk through skin cancers and related diseases. Despite scientist's warnings, we continued to use gases which broke up our atmosphere until our only option left is to leave this dying planet and search for another. Yet the government in all its infinite wisdom wants to use money destined for space research, to purchase nuclear weapons to assault our enemies who we no longer have.

"Some people even oppose space exploration; they are so stupid they don't wish to save their own skin or even their children's. That is how naive we have become, so now God is putting an end to it, cleaning up what we have destroyed.

"As I speak to you trees are blooming in the church courtyard and flowers are opening their petals as if it were the first day of spring. What a spring they are entering into, they are welcoming the second Messiah. For those of us who have different religions, I do not wish to spoil your day, but very soon Mary is going to say who you have to worship. It will then be up to you to make your own decision, do you believe in Mary and her Mother God, or do you believe in something else; the discussion will be yours to make.

"It is very important to mention at this point, although this Christian church has been selected for the birth, although I doubt if Mary will be born unto us like Jesus was, even Christians may have it wrong, who knows? With the changing of God's sex, the new bible and so forth, Christians could well be wrong in worshiping God, as God wishes us to worship Her."

Tony stopped suddenly, stunned at what he had just said. He couldn't believe his own ears, yet he was saying it and he couldn't stop himself. For once, in his entire epilogue of commentating over the subject, the producers and managers were glad, even smiling, overjoyed, that he had said those very words. Yes, Christians could well have it wrong, something First Lady Rocc had not thought of. . . Yet; if she was wrong, it would be a terrible shock for her and the church.

A short time later after regaining his thoughts, Tony started his commentary again, picking up from where he left off.

"No doubt Mary will lay down laws and give us guidance in a religion many of us have either forgotten or cast aside in our efforts to please our parents for our own self gratification. We have forgotten why we're here; not questioned the word of God and asked for Her forgiveness; I'm sure God forgives us all, sinners alike; let's hope so anyway, for if not, this must be the Day Of Reckoning. The day we have all feared, when the dead will rise and life will be immortal; nothing will be as it was before."

He stopped as a soundman gave him a signal something was happening behind him. He turned around slowly, giving his two cameramen a chance to see what it was. To allow the people at home and around the world to see the church doors open and hopefully, the coming of Mary.

The ground was cleaned intensely, instead of cobblestones, pure white marble, brilliant white, was shining brightly, despite the sun was in the east and was not yet bright enough to cast shadows or make materials sparkling white. It was so bright; the lens on one of the cameras had to be shut down to their lowest stops and filters added to see what was happening before the white church doors.

"I hope you can all see this ladies and gentlemen; God is cleaning the floor of the churchyard to the entrance of the church, turning cobblestones to pure white marble. It's extremely white, so bright to us that we have had to add filters to our cameras to stop the glare damaging the lenses and intricate parts which send the send the pictures to you.

"It is not only our cameras, but every camera here has been affected by this ultra bright whiteness." He paused; allowing the cameras to pan around the area, thinking of what else he could say to explain how he felt at this precise moment and what was happening.

From the cobblestones, the new white marble suddenly expanded, like a flash of light along the black tarmac path which led to the road. There were two paths, both crowded with press and camera crews. There was another path which surrounded the church, allowing the vicar safe access from the main doors to his house. The flash leapt from the cobblestone beneath the press people's feet as if they were not there; then it went around the church to the main car park, no one could speak for a moment as they were all stunned. A few moments later, Tony and the other commentators started speaking again.

"Did you see that? The light just shot out without warning; right under our very own feet. The path we are standing on is completely level and white, despite our black muddy shoes; the path doesn't mark at all. It's white and I have a feeling that somehow, it is going to stay white, no matter what is put on it.

"I am just waiting for confirmation as we have sent a runner around the church to see what is happening there. When the white light came under us, it followed a different route; we were not injured and didn't feel a thing as the light passed below our feet. It was the most thrilling, exciting thing which has ever happened to me. I also speak for my crew, nothing like this has happened to the others either. We feel on top of the world, hold on, here is the runner and he's nodding his head and calling out. . .Yes," he said, putting his hand to his earpiece.

"The path is completely white right around the church and to the car park. Each path has been changed to white marble up to the outside perimeter and pavements. The old gates to the church and churchyard which I noticed upon entering the church were hanging off their hinges, now they have been replaced and are a brilliant pure white metal. They are hanging correctly aligned, apparently fastening themselves when the gates were left wide open.

"It's a miracle all right; there is no other word for it which can explain what we have just experienced, this had to be done by God, do you honestly think Lucifer, the person we have come to know as an evil God would wish to surround a church with such brilliance and splendour? Would He wish to cleanse the church and the people before they enter it?

"Only God could do this and it's definitely a woman's touch; a man just wouldn't think of doing this; the building yes, but brilliance of this kind, beauty such as this, not on your life. This has definitely been done by the hand of a woman," Tony sighed, knowing he had told the truth and nothing but the truth; as God had wished him to tell Her people.

He had said what was needed and managed to get his point across to the people. He had proved beyond all doubt to the people at home, to the women at home, who would have the influence needed to pressure their husbands into believing God was female. God was more powerful than the Devil and Mark was no more than the Devil's emissary who would in the end, loose to a greater power, greater loved and greater needed. . . God.

"Good morning to those of you who have just joined us," Tony said quickly. "It's now six twenty five, the expected coming of Mary on the dot of six didn't happen. However, at six precisely this morning, miracles did happen.

"Our atmosphere has been cleansed; trees and flowers are in full bloom; birds are singing their hearts out, praising God on high; thousands of Robins are gathered in the churchyard to witness the second coming. They are still here, and are now being joined by other families of birds, even Golden Eagles are sitting next to the much smaller Robins, as if they have been friends all their lives. Some of these birds normally migrate to the south this time of year; blue tits, wrens, swallows, swifts and high in the air, circling above us, the greatest array of birds you could wish to imagine, all waiting to see the arrival of Mary.

"Perhaps the most spectacular and to us standing here, the most exhilarating moment was the change of the forecourt from cobblestone to white marble. We can show you this moment again," he paused, looking at his manager who was looking to a marquee tent which had been erected the previous day. This was for the TV crews to use to get their programmes out; a moment later, Tony had a signal and was told through his earpiece everything was ready.

"Yes, we are ready." He looked at his own monitor and talked the people once again through the spectacular change of black tarmac and cobblestone to white marble which refused to mark in any way.

"The doors of the church were changed yesterday to a gleaming white and been purified by God's Holy hand, but now they look even brighter than before. How do I know this to be true? First Lady Rocc, who is standing no more than eight metres away, has just told everyone here by thought what has happened has been done by God. That too is amazing, until it happens to you, you don't know how brilliant thought transfer or telepathy, even one way telepathy really is," he said joyfully. He watched the end of the tape play and turned to face the church doors.

"This entire procedure has taken twenty five minutes, and now it's almost six thirty, six, twenty nine to be precise. No one knows when Mary will come, although with the gathering of the birds, the cleanup of the atmosphere and the cleansing of the paths, I'm sure it won't be too long now," Tony said reassuring everyone.

"Tony, fill in for a few minutes, we seem to be picking up rather strange readings on the equipment," Malcombe, his technician said through his earpiece.

Tony continued automatically going back over what he had previously said before like any good commentator would do. "If you were up at the crack of dawn, around six, you may have noticed the most fantastic sunrise we have ever witnessed here in the United Kingdom; it was just like being on a tropical island by the equator," he said thinking quickly.

Then he noticed the church doors, they were starting to open and the group of disciples moved, forming two lines with First Lady Rocc standing between them, in line with the doors at the head her the disciples.

"It's happening!" Tony whispered into his microphone. "I am stood directly in line with the church doors; Paul, my current cameraman, is stood directly behind me and what I can see, you will see too," he said quietly, forcing the sound technicians to turn his microphone up for transmission purposes.

He gasped in awe and wonder. "It's about to happen and you people at home, or wherever you are, are about to witness the most spectacular event since the dawn of our planet; this is going to be better than the birth of Jesus Christ."

He stopped, mouth agape, gazing hypnotically at the church doors as an eerie silence fell over the assembled crowd, while the church doors silently opened before them.

CHAPTER 12

All cameras and eyes were now looking at the church doors which were wide open; a bright gold cross, suspended in the air above the entrance moved forward by itself. It slowly moved out of the doors and continued its journey rising up to the front of the church until it was high in the air and the bells of the church rang out in praise in perfect pitch and harmony.

Along the back wall, Mark looked up from his new perch on the scaffolding and muttered words in a very strange language. People standing close to him could hear him, but not understand what he was saying.

As they faced Mark they looked deep into his jet black eyes, seeing the depth of space and the empty region within him. Stars were twinkling brightly through his black velvet veil, calling people to enter his domain, join with him and put down the word of God. He held up his hands facing the people standing below him and behind the wall.

"You people gathered here this morning, have you lost faith in what your hearts' desire so much? Have you lost the satisfaction of wealth and prosperity? What is the matter with you people? Why did you come here to see a show of sin? This is not God you seek, but some other aspect.

"This is not wealth; this is not what you crave for; this is not what you have worked hard for, you deserve what is yours by right and I can help you all obtain more wealth than you could imagine and know what to do with. You will all be very happy in my Father's Kingdom; a Kingdom He has made, right here on Earth for all of you."

He emphasised his words well, trying to make people realise just what they had and what other riches lay ahead. All they had to do was turn away from this false God and join Mark.

"Are you really taken in by what you see; a lot of mumbo jumbo, clean air? In time you would have made the equipment to clean the air yourselves, it's not as if you are not well onto the way to controlling the weather and I guarantee you, within the next two years, you will have the technology to accomplish this yourselves," he roared.

Mark glared at the people before him, now they were taking an interest in him. The thought of wealth had drawn them from the coming of Mary.

"Wars, well, wars help everyone realise there is more to life than social gain; there is territory gain as well, getting more land, you fight wars every day of the week."

The people around him looked mystified; there certainly were no civil wars, or any other world war and the UK was definitely not at war within its own borders.

"I see you doubt my words, tell me if you disagree; don't you buy land, argue among solicitors to gain it? Pay more for it, barter for goods? Do you not try and knock down the price to get a better deal?" He could see the people gathered close to him nodding their heads in approval.

"This too is war; it is not fought with weapons and bloodshed, but money and power. The more power you have, the more land you can get; the better deals you can make. People in poor countries don't have money and your type of power in a society which has grown above barbarism. However, in the countries with no money, the only thing they have to barter with are the lives of people; by the millions if necessary. People are willing to sacrifice their lives to better their children's future; give them a better chance to grow food on richer pasture; have that extra square kilometre of land to expand on.

"Just like you wish to expand your four bedroom semi-detached house into the garden next door; you barter to buy their land and if you pay enough money, then you will get it, but it didn't cost your life, just a few thousand pounds, or pieces of printed paper. So you see, there is war over the entire world; it is not so bad when you break it down like this, to its fundamental units of currency and wealth," he shouted into their minds, laughing haughtily and glaring at everyone.

"What about the way God has changed our health? She made us well, our correct weight for height," a woman shouted from the back of the crowd.

A group of people standing close to her glared, as if they meant her harm; then they shuffled forward, back and both sides, to completely hem the frightened woman in. It was like they were trying to suffocate her small area of space.

"A very good question, but now I ask you, did you ask to be put like that?"

"No!" a man shouted in reply and looked to his neighbour, giving him a menacing look. The next man agreed with him and shouted back "No!" and the next and the next and so the no's continued, all through fear.

In truth, they felt better, able to move and breathe than when they were overweight. However, with the thought they would be hit or killed, they decided to be quiet against their better judgement.

"So you see, a majority of you don't wish to be healthy and your correct weight. I bet, if I offered you free; a hundred cigarettes a day for the rest of your lives, you would take them and smoke the lot. In fact, I'll go further than that," he paused, gaining everyone's attention.

The small crowd were now enthralled with his outlandish proposals; nobody had ever made offers like this before.

"I would like to offer you people here," he took another short pause while he looked around the eager crowd of faces. "You can have as many cigarettes, tobacco of any form, whether for rolling your own or pipes, you could possibly smoke a day for as long as you live." The crowd was stunned and very silent.

"I would like to take you up on your offer," snarled the same man who had started the procedure from the questions before breaking the silence. He looked around to others with glaring eyes and fists that threatened others close to him. He urged others to speak up, join him, as he punched the closest man in the face, making his nose and lip bleed, he threatened others as well and before long, half the people in the crowd had agreed.

"Then here you are," shouted Mark.

From nowhere, thousands of packs of two hundred cigarettes dropped from the air into reaching hands from the crowd. They came complete with lighters and matches as people preferred. As the people reached out with their hands to catch the packets, they turned to flowers in full bloom, smelling of the sweetest perfumes anyone could imagine. The instant craving they had been given was forgotten, pushed to the back of their minds.

"You will not win my people over that easy Lucifer's child. You have to grow and cherish this world to understand the people who live in it. Do not come here expecting to change people's minds with your hasty offers, you will discover kind deeds do not always win the hearts of these people, they are hard, well worn through the passing of time," a female voice boomed through the air over the crowd, dominating their attention. The man with the bloody face was suddenly healed, the man who had punched him, found he was no longer a tough man, but had small hands and small muscles. Everyone knew from instinct it was the voice of God, a voice they had not heard out loud in thousands of years since the time of Moses, She had remained silent.

"You people, listen to me," God called aloud, making everyone hear, even the deaf and partially deaf could now hear, some for the first time since their birth.

Spreading Her voice over the world, She ensured the deaf could hear and blind could see. Not in just one or two places, but people everywhere.

"Neither Lucifer nor Mark shall ruin My daughters' day, it is her time to come forward and for to you to listen as she speaks on My behalf, for you to judge as you have been allowed in the past to judge between good and evil. I have never taken that away from you since the Garden of Eden when I freed you into the real world."

"Please God," said a shy woman from the middle of the crowd. "If you please, I know there will be many questions answered in the coming months and possibly many changes to our lives. But . . ." she paused, wondering if God would continue to listen to her petite voice.

"If the Garden of Eden was once in existence, where was it?" she asked.

The woman, now silent, was frightened she would be punished by a bolt of lightning for interrupting God Almighty in Her speech. As she looked to the skies for the damming lightning bolt, God's calm, quiet voice spoke, not harsh, but in a time of remembrance; as if recalling the day, many thousands of years ago when woman and man walked alone and naked in fear of nothing. Having everything they desired, food warmth, friendship, but not having the pain of love, fear, want and children; to know the difference between life and death, to know the difference between love and hate, to kill and be killed, to have to work or die for a crust of bread to survive.

"The Garden of Eden existed a very long time ago. After I had created the heavens and Earth, the argument between Lucifer and I started and finally finished. I watched the people of Earth flourish and become lonely, therefore, right or wrong, I made two people for a test; first Eve, second Adam, I put them in a mythical garden which I called the Garden of Eden which was for them alone, it's actual location? Well, it's not on Earth," she tittered. "Neither is it in your dimension of space; it lies between my Kingdom and Earth.

"It is well guarded by the Flaming Sword, flashing back and forth, it keeps everything out and to see it, you must be pure of mind and spirit and heart. You have to learn to travel vast distances through space and through Black Holes, the keyway to different galaxies. Black Holes are very powerful, they need to be to fold space and that is exactly what they do. One or two of them, and I will not say which ones, but they will take you right to its location, but that is all that you will see, the entrance, for it is so dear to me, it holds a treasure I have never allowed anyone or anything to enter unto it.

"Eve and Adam remained my companions for many thousands of years while life on your planet grew from a few too many. They listened to stories I would tell them and I too listened to them. We had a wonderful relationship and they were my children; I placed the Garden of Eden away from Earth, away from sin and horror which Lucifer had saw fit to give the humans. In turn for their possessions, I gave them life and death, a way to leave their planet and join again with me to be in tranquillity where Lucifer, no matter how hard he tried, could not prevail; My Love and My Kingdom.

"The world he took from me, He turned it from good to a violent, desolate world because of the hatred He had for me. After He penetrated the Garden of Eden and took my children from it, I then banished them to live on His desolate world and live like the humans. They hated me for it, but there was nothing else I could do.

"The Garden of Eden was not meant for evil, the things you now take for granted; it is like being in Heaven itself, so pure and kind." The calm voice died silently away for a moment or two and everyone wondered if She would speak to them again.

"Mark!" It suddenly exclaimed seeming to be directly over his head. "Be gone for the rest of this day, do not show yourself to any man, woman or beast, for if you do, I will bury you deep inside Hades to be with your Father forever more!"

A brilliant, electric blue, majestic bolt of lightning starting high up in the stratosphere, hurtled down upon the group of people like a rocket out of control. It charged once, charged twice and charged again and again with the collective force of static electricity gathered in the entire Earth's atmosphere.

Mark looked up on hearing a loud whistling sound above his head but was surrounded by a group of women and children he had used to protect Him from His Mother. The lightning bolt hit his head, travelling straight through his body into the very depths of damnation itself so even Lucifer felt God's wrath.

Before everyone's eyes, Mark was transformed from the living man he was into a black slithering serpent which slithered down and down, ever deeper down into the endless pit the lightning bolt had created. Yet for all its great collective power and force, only Mark was affected; not a hair was harmed of anyone else, not even the people standing next to him and touching his clothes.

When he was gone, the pit closed and the air was so sweet with perfume, it would have taken a trillion flowers to perfect. The people were silent, wondering if they too had offended God by listening to Mark.

"Go my children, enjoy this special day, for there will never be another like it until the day you die and join with me in My Kingdom." The voice was gone and everyone knew it would not speak again.

One of the most puzzling questions throughout time had been boldly asked by the meekest of people and thankfully answered. The world's population knew the answer to the Garden of Eden. Historians would finally be able to tell the truth about it, everyone was sure; more truth than ever possible believed would soon flow from the lips of Mary.

In the churchyard, while God had spoken and the world had listened, more and more flowers had grown and the grounds were now more splendid than any flower garden ever created by man or woman. In all the history of man, all the times man had tried to perfect gardens, none could match this day's beauty and fragrance.

During this time, a low humming of angel voices could be heard from within the church itself, but none dared enter the church without first being invited in by God. They expected to see Mary outside, like a blushing bride, whatever was happening inside, was between God and Her daughter alone.

The sun, with its huge mass and weight, raced higher up through the morning sky, going through its daily routine but now travelling much faster than ever before, being driven by the Hand of God.

By 07.00 the sun was at its zenith, standing still in space and there it remained, almost threatening in its nature, casting no shadow over the church. Standing directly over it, out of its normal line of ascension, as if a warning to anyone who dared venture in uninvited, or any misuse or miss conduct happen. The sun would be there to guard the coming of Mary, in a threatening grand gesture sending fire and brimstone at the person, persons or group who dared go against the word of God on this, Her special day. For two thousand years, God had rejoiced as She was so doing now. A gift to the people of Earth from God would never happen again.

The sound of angels singing low in the church grew louder and louder until it joined in the throng with the bells, rejoicing in God's Almighty name. At the same moment, even in churches where only records were used instead of bells, bells pealed throughout England, Scotland and Wales and the whole of Ireland. The bells also sounded throughout Europe, Russia, India, China and through America and the western world. Church leaders were stunned as bells sounded without their intervention, and many churches had bells sound where they had never been given bells, but they were now installed in the church towers, which were suddenly added by God's hand to those poorer churches.

Throughout Australia, New Zealand, the South Polar communities, all islands in the southern hemisphere, up to America, north to Alaska and the Arctic Circle and down the other side to join Russia. Nobody could now say this was all put on, they could see for themselves, hear the bells toll where no bells had ever been before.

Many church ministers said this could not be happening in their cults and beliefs, they were afraid, annoyed with God, in all the countries of the world, God had picked a small country called the United Kingdom for Her gift to the world.

Yet as they listened to the reports of why the UK had been chosen, for its non use of weapons to keep the peace, and its court system; the church in all its beliefs, indeed even the Pope was not seen on his balcony while he listened intently to the broadcast in his own language of what God was saying to the world. God and the Anti-Christ Mark were here on Earth, together for the first time in they didn't know how long and the Pope was devastated, to think the beliefs he stood for, did not truly represent those of the true God.

Churches in South America rang their bells but there was nobody there to pull on the ropes; they tolled by themselves, calling people from their homes to the church to worship God. Further north to Canada, where again the church bells rang, despite the religion of the church, the bells rang just as loud and for the same amount of time as everywhere else.

Throughout the world, each different belief was worried theirs was not the right one, yet through all the talks and discussions, not one single word had been spoken about the belief by God. As the High Priests discussed matters between themselves, they found it true; God had not condemned one belief: at the moment.

Some people questioned if every belief was right, but those questions and many more could not be answered by mortal man, they would have to wait until God or Mary decided to inform the world what belief they were to follow and how.

One thing was certain, in every church around the world, in every persons hand; the new Green Bible had just been given. No church would make money from God's own words again; the bibles were free to everyone, and everyone would have a copy of their own bible, even babies. The bible would answer most of the questions people had, the rest of the answers would come very soon.

The bible told of other life forms which populated the universe and grew in different galaxies. It said that women, not necessary man would have the opportunity to visit and preach to those species.

As the bells tolled, rocket designers in America, Russia, Europe, the UK, China, Japan and India, were coming up with ideas of what the new type of space ships would look like and how they would travel the vast distances of space; they now understood space like they had never understood it before. What had upset the directors of those companies most was the designers were all women, a supposedly inferior species, only good for reproducing their life forms.

Women were coming up with the answers which man had been searching for, for the past fifty years; it was now taking their colleagues, men, time to understand what the drawings and designs were about. Men were even suggesting their ideas were unfounded by mathematics and the ships would never get off the ground, let alone fly through space to other planets thousands of light years away.

As the bells tolled, the women of these designs, worked faster and faster, communicating by telepathy on their designs; as if being drawn by the sound of the bells into some form of trance that combined all their minds together to act as one, making one very large team of designers.

An hour later at 08.00, the bells stopped ringing right around the world at the exact same moment. It was then the women rocket designers put down their pens and looked up to Heaven, for they knew in their heart's, Heaven was indeed up; among the stars, separated from them by dimensions which could only be reached upon death itself.

Their first designs were completed, their work for the moment finished, now it was the time of God. For God to place Her daughter into the hands of women and men, this time, Her disciples would be there to ensure no harm would come to her. She would not be betrayed, she would not be turned away from the church or synagogue alike.

She would be allowed to roam the Earth as she pleased, God would ensure this would not end in the catastrophe which had happened before. Leaving men confused and unsure and leaving the men to do God's work. The first plan had failed to allow Jesus to carry on His work until an old age, but the name of God was still here two thousand years later and respected around the world, so it had not completely failed; but at the same time, there were now numerous faiths, whereas there should only have been one. However the rules this time would change and God would remain supreme for the remainder of women's days.

At Radstock Church, when the bells stopped ringing the inner doors to the church opened revealing colour and light, brighter than man could ever imagine. There was so much radiance, for a brief, moment, the group of disciples and First Lady Rocc had to cover their eyes.

It was beauty beyond beauty, a police officer, dedicated to his job, was only interested in seeing if there was any damage inside the church. He had convinced himself this was all a charade, put on by someone who wished to cover up the chaos which would follow and the horrendous damage done to the inside of the church.

The mess inside the church, the stolen crosses and gold would be enough to pay for this scene several times over. He stepped forward, up to and level with First Lady Rocc who turned to glare at him, but he shut his mind to her and managed to block her thoughts.

First Lady Rocc knew immediately this was no ordinary man; he had been recruited by Mark and given some power to get by them and upset the mornings' events. He was planning on upsetting the world's second coming as Mark had attempted to do himself. She put her hand up to stop him and voices called, begging him not to enter the church but he stepped forward, pushing all the time past the disciples with great force. He was definitely not himself as his eyes showed he was determined at all costs to enter the church.

A single thought entered the group's minds and they stepped back giving the man a clear path to the church. He stopped, then stepped forward and stopped again, getting closer and closer to the main doors.

"STOP!" First Lady Rocc shouted, almost deafening the nearby audience and camera crews.

He did not however obey her command; he continued forward, each step in slow motion, getting closer and closer to the main doors.

Then it happened, as God had foreseen it would. Lucifer would have many tricks ready to impart on Her great day, to try and show Her up and put Her down by the people of the planet Earth, Her own children. A beam of continuous light shot down from high above, where the sun was watching the events with an assurance it would save the day. The light was pure, although it burned the eyes of the people who dared to look at it for too long.

The man stood still, stationary in mid step, bathed in the sunbeam which in the song of old, children sang so often in Sunday Schools. Here was that sunbeam, holding this man of Lucifer and changing his mind, burning out Lucifer's thoughts and controlling power.

"Is God going to burn him to death?" Tony asked Victor his soundman standing close to him.

"Heck if I know," was his reply as he concentrated as best he could on the situation no more than five metres away.

The man remained in the sunbeam for an entire minute, then the beam was gone and the man completed his next step and shook his head as if trying to understand what had happened. He stopped, and slowly turned around then fell to his knees, his hands together in prayer before him praying to God for forgiveness.

"God has forgiven you," First Lady Rocc said, placing her hands on his head and also praying to God for the man's salvation. "It is time for you to return to normal duties, you are a cleansed man. Have faith in God and She will have faith in you."

"How did I get here?" he whispered, looking at the vast crowd before him, blushing profusely. "I recall nothing except a stranger called Mark."

"That was Lucifer's son, He has been banished from Earth for the rest of the day and now you are free from his control. Go and join your people, there is nothing to fear God has cleansed you and will take care of you forever," First Lady Rocc replied soothingly.

Slowly, she led the confused and shattered man away to the side where he was engulfed with friends who took his hands in admiration for what God had just saved him from.

"This is truly a morning of miracles," Superintendent David Prentice said to him, smiling and shaking his hand in approval.

Then a thought entered his head, as too did it enter all the all the minds of the disciples. If Mark had recruited one man, then others would surely follow.

"I know this is an auspicious day, but it is going to be recorded on film forever," said Superintendent David Prentice to his group of officers. "I will ensure you all get a chance to see it later but," he paused and walked among his officers; searching each person's eyes deeply. He took a deep breath, looked across the crowd of people and realised any one of them could be the next person to try and wreck the day.

"Nigel," he said to the officer who had just been saved by God. "Go to First Lady Rocc; say I wish to communicate with her from where she is: then return to me." Nigel immediately set off on his short journey.

"What is you wish David?" a voice said into his thoughts.

"How do I talk to you?" he asked aloud to no one in particular. His officers looked at him confused.

"Just think your question, there is no need to speak the words, you only show yourself up if you do," First Lady Rocc replied in thought.

David blushed slightly and turned away from his men facing the crowd, looking at them as if he were lost in thought for a moment, thinking on his next words.

"What happened to PC Nigel Hayes made me think, my men can search among the crowd for another imposter."

"Imposter?" questioned First Lady Roc.

"I mean one of Mark's dark angel's so to speak," he thought to her, not really understanding how telepathy worked, but was happy nobody could hear his outlandish idea.

"I see."

"Before I start, I would like you to please check all my officers are not contaminated by Mark's thoughts if you see what I men."

"I understand what you mean," she paused for a brief moment. "It is done, all your officers are clean, it is a very good idea, however, I can also tell you this has been in our thoughts as well. Between us, we have checked all the press and people near to us, but we have found on the other side of the wall two people who we cannot read. They look like this." A picture of both people appeared in his head and the same pictures were sent to all his officers in his group.

"You may take them to one side and separate them from the crowd and I am sure God in all good time will deal with them in the way She sees fit."

"Thank you for your help, we will get onto this straight away," he replied.

"I have just been informed the people in the area below the wall are now clear. More people are arriving above the wall and we have not had time to check them. The main ceremony is about to commence, perhaps you would like to start on the new people and check them out for us?"

"Is there anything we should look for in particular, to know if they are one of Mark's dark angels?" David asked.

"Their eyes will be jet black and so too their hair, it will be as black as coal," First Lady Rocc replied and left his thoughts as lightly as she had entered his mind.

David turned to his face his officers. "You all know the two men we are looking for, be very careful they could be dangerous but use no more force than necessary, it will do no good to cuff them as they are very strong and will pull the cuffs apart and then they might try to harm you. Speak to them calmly and request they follow you. Take them to the holding area by the communications van."

Everyone looked at him amazed then stood to attention and left to carry out their orders. They didn't have to be told the remainder of their duties; they seemed to know what they were as it was all passed onto them by First Lady Rocc at the same time she spoke to David.

When the two dark angels were approached, they knew what was about to happen; they fought with all their might, using the gifts which Mark had given them. People standing close by, diverted their eyes from the large screens to watch the police try and capture the two men which was turning out to be a futile attempt. The two dark angels were strong in mind control, much stronger than Nigel, who had been gifted by God with limited power to help him during the day. It was turning out that Mark had spent more time with these two men than anyone thought.

However, when Nigel managed to get through the ranks of police officers, the two dark angels looked at him with trepidation in their eyes. There was something else about this man which was not in the others. Nigel stepped forward with arms outstretched; his hands open wide. The other officers did their best to hold the two dark angels as Nigel touched each man on the head.

Instantly they calmed and their hair returned to its former colour, changing from jet black, which Mark had given them to show their allegiance to Him. Their eyes then turned from jet black to blue, an electric blue, the same colour eyes which Nigel and the disciples now had. The two men finally calmed down and stopped wriggling and shouting strange words in a language long forgotten by man; cursing God on High for all they were worth.

In a few minutes the two men were calm and peaceful, shaking their heads, asking question upon question of how they got here.

"David," First lady Rocc called to the superintendent.

"Yes?" he replied aloud, then realised it was First Lady Rocc speaking to him. He was still not used to being spoken to in this manner. His officers turned to look at him, then realised he was talking to a disciple.

"May I suggest you take the men away to a safe area and question them? Find out what method was applied to them by Mark and where they were when it happened, if they can recall these details, I think this may come in useful when you find others."

"Of course First Lady Rocc," he replied.

"Please, don't be rough with them, they were not in control of their actions, you are dealing with a power much higher than any other on Earth. One other thing, do not bring any charges against them, or any others you find throughout the day, just get as much information from them as you can and then let them go. Try giving them sweet coffee and chocolate biscuits, this will assist in helping them to recover their memory, and this is what you need. Do not put them into prison cells, there is no need for this any longer, God will rule us now and I am sure things will change for the better very soon." The conversation ended and David went with the two men and two of his officers to the holding area.

"While I am gone, start the search for any other dark angels; Nigel, you seem to be able to control these people, you are in charge. While I was talking with First Lady Rocc I had the feeling we are being scrutinised by people on our left, I suggest you start there."

"Yes Sir," Nigel said. The others stood to attention for a brief moment, showing they had understood and would comply with the order.

On the large screens above the police officers who were doing a desperate job to make it a really special day for all, the church was becoming full of colour. It was brightening up like a rising sun; the sound of angels was now singing praises as their voices filled the church below the cross.

Something was happening inside the aisles, it was difficult to see, no matter how television crews tried to peer inside, their camera lenses, even telephoto lenses, couldn't get past the main doors. God had forbid it, no man; woman or child would see the inner sanctums of Her church until She was ready.

The crowd of reporters and journalists gasped as they saw the first movement from within the church. Whoever it was, he was making their way slowly down the aisle; the person was dressed in white with a gold halo over her head. The people outside could just make out the form of a woman which had to be Mary; soon the press and the world would see her in all her glory.

CHAPTER 13

The choir of angels grew louder and closer as Mary neared the main doors. Now the crowd could see her on the huge screens outside. She looked magnificent, radiant, beauty beyond beauty, definitely feminine, as all women should be; even at this distance her form could clearly be seen beneath her pure white satin gown, gathered with a simple gold cord.

A brilliant gold halo hovered a few centimetres above her head, fluctuating through the colours of the rainbow, showing splendour and glory. She walked steady and firm, her shoes, low healed, brilliant white; shining like her dress. She was at present not a woman of up to date fashion, she would be setting no new trends for women to follow, in rich exotic gowns and clothes in different materials, far too expensive for the young women and girls to purchase.

She appeared to be in her late twenties, no older than thirty, so she should be able to make the young people want to follow her around and copy the way she dressed, but in her current clothes, none of the fashion reporters thought the shops would be selling this type of robe.

The fashion journalists, whose job was covering the clothes she wore, were hoping to get an interview with her of the type of clothes they wore in Heaven. Now, as cameras clicked furiously, they were disappointed in what they saw. They had expected the daughter of God to appear in the most feminine of feminine clothes, a clinging dress showing off her curves, long golden legs, high heels, a real stunning woman with looks to die for, hair golden in a style out of this world making every woman envious of her and wanting to be like her and follow in the clothes she wore. Apparently, Mary didn't wear makeup, she didn't need to, her complexion was young, clean, pure and soft, matching the single white satin robe she wore. Makeup would have only made her look worse and as the cameras reached her face and zoomed in on her eyes, they could see her eyes were an electric bright blue colour, then someone realised and told everyone that her eyes were the same colour as the disciples.

Some of the women reporters for Hair, Cosmopolitan, She, Fashion Scene, Endore, Makeup and Beauty, decided they could write nothing about this woman of beauty or elegance. She obviously had no sense of dress compared to modern day women; she wore no makeup, not even the slightest hint of lip gloss or eye liner.

They turned away in disgust, shutting down their IPads and tablets, pushing tape machines into their pockets and handbags and pushed, in an angry mood, back through the crowd, dragging their male photographers behind them, cursing aloud as they walked away in disgust.

Before their machines had time to shut down, they started up again and appeared back in their hands, a woman's face filled their minds. A woman of beauty, so beautiful that she didn't need makeup, she wore sensible clothes, they didn't cling, hug, embrace her body; her skin could breathe the clean air. Mary was wearing clothes for her and no one else; she was not a sex symbol that was clear.

Her skin was so soft and gentle, it appeared as young as a baby's and makeup they all knew did not befit a child, clogging its pores and stopping its face from breathing, evolving into a woman. It looked ridiculous at times, sometimes horrendous and grossly out of place. Makeup was indeed a sign for a show of sex and Mary was not into sex; taking a mate from

Earth was not going to happen. Men would fall at her feet to walk by her side, but Mary was not looking for a man to have at her side, she wasn't searching for a multi millionaire either, Mary had all the riches and power in this world and the next. A mate for her was out of the question, no Earthly man could dare shape up to her expectations, no man on Earth could be that pure.

They suddenly realised this woman was a real woman to write about, her clothes were simple, her shoes would perhaps catch on after all, in their own way, Mary could start a new trend in fashion shoes and even makeup, despite her lack of it, through her pure body alone. Women all over the world would want to have a face, skin like hers and cosmetic firms would soon be putting creams on the market to make the face as young and soft as baby's skin.

She was the perfect height, two metres exactly, so she could be seen over the heads of people and would stand out in a crowd; she had the perfect shape, not a millimetre of flab on her body. She was a woman of sheer beauty; she should be seen and spoken about, her pictures in every magazine in the world. Every paper would carry something and their story had to be the best. It would be! They had to tell women all over the world what they stood for.

Their new rights, it was now time for men to chase the women in all fields, men should now look up to them, for they were now above men and would be forever more. They took out their tape recorders from their bags and returned to their places with their tablets ready to go. Confused photographers were getting their cameras out again and the women were telling them what pictures to take and how and to take them, even what lenses to use. They started to write, describing in detail every attribute to the Lady of the Day.

Mary waited by the new church doors, leading from the aisle to the inner church hall. Something was happening, or about to happen, no one knew what.

Another figure, thin, almost transparent at first floated down into the church hall, nobody could be sure what really happened. The figure, most likely an angel, materialised inside the main church hall; as the people outside watched through long telephoto lenses, which sent pictures to the large screens overhead, the ghostly spirit grew more dense, gaining more substance and formed before everyone's eyes.

The angle had its back to the people outside and was facing Mary. The crowd could only guess it was a last message from her Mother, giving her last minute guidance as to what to do and how to do it. When the angel finally turned to face the crowd, it too was dressed in a white satin gown which glowed. Now the crowd could clearly make out the white wings on its back, a mark of the Archangel Gabrielle. Gabrielle spoke softly but firmly to Mary, something the crowd had not thought about. This was the house of God; this ceremony would be different from that of Jesus' birth.

"I am the Archangel Gabriel, I will be leading the ceremony from here on in God's Holy name," he said to the people gathered outside and on the streets throughout the world. Despite the fact that he spoke quietly, everyone heard his every word as he spoke directly into their heads.

"The choir will continue to sing praises to God and pray for the next five minutes. Then we will have the main prayers and the Daughter of God will finally leave her sanctuary and set foot upon the Earth. So that everyone is clear, Radstock Church is under the full protection of

God and is technically on the other side, in Heaven; you are allowed to see it but Mary is not yet on the Earth and neither are we. This is what you people call, a hologram."

As everyone looked at the church, behind it was another planet, it was very large, yellow, with blue and green areas which could be seas and green land. In the upper right hand side of the planet was a huge cylinder floating in space, very wide and very long, and inside it, could be seen what appeared to be hundreds if not thousands of galaxies, stacked on top of each other, glittering in the darkness of space behind it. Then a spaceship, very long and wide, passed behind the cylinder and was out of sight for a moment before the ship came back into view then moved off.

"This is a view of another stellar system, one which is very dear to us and will, at some time in the future be shown to a number of you, if you get into space. Have no fear, the people on that ship are like us, angels, and we too like to explore God's Heavens and see Her wonders, it is not all doom and gloom in the hereafter," the Archangel Gabriel said to everyone.

The view of space disappeared as the church once again took up the full view of the hologram. Gabriel looked concerned for a moment that something had been seen that should not have been shown to the people of Earth, but now it had been done and there was no going back. Knowing he had not lied to the people before him, he knew it would make them think and want to move into space to discover where the scene was, so he moved on with the service.

The choir of angels started to sing its praises to God; which was magnificent to the human ear. The Earth audience was captivated with the signing, it wasn't what everyone was expecting, but music to soothe and captivate everyone's ear.

After the songs, Archangel Gabriel stepped forward to the main doors, raised his giant wings behind his back and the angelic music and song died away. He sang the following prayer, with a huge choir of angels backing him.

The prayer was known to many people of Earth too, but as they started to join in, they stopped, for many of the words had been changed. The prayer was in a different form and this was the very first time it had been sung and led by an angel in their presence.

Be still for the presence of God,
Her Holy one is here.
Come bow before Her now,
With reverence and fear.
In Her no sin is found
We stand on Holy ground
Be still, for the presence of God,
The Holy One is here.

Be still for the Glory of God
Is shining all around
She burns with bold fire, with splendour
She is crowned.
How awesome is the sight, our radiant
Queen of light!
Be still, for the Glory of God is
Shining all around.

Be still, for the power of God is
Moving in this place.
She comes to cleanse and heal, to minister
Her grace.
No work too hard for Her,
In faith receive from Her.
Be still, for the presence of God
Is moving in this place.

"All praise be to God," Archangel Gabriel said and everyone was silent.

The churchyard and local area was completely silent, you could have heard a pin drop onto the new marble floor. Not a voice could be heard, not even the birds sang their songs of praise, not an infant cried, and not a soul dare speak out of place. Mary was about to step foot upon the Earth itself and no one dared wish to upset the wrath of God, least he or she displease Her and would then feel Her mighty hand of fete.

Even the angels were silent, Archangel Gabriel in all his splendour, did not outshine the woman stepping from Heaven onto the Earth, from the inner chamber of the church, forward to the main church doors. Everyone suddenly noticed a distinct smell of perfume, pleasing to the nose; cover the area surrounding the church.

It was as if she was not breathing, for those stood before her bowed low as Mary stepped forward and looked about outside of her domain. It was as if she were an infant, about to be born into the world and not sure whether to leave the safety of her Mother's womb or not. A gentle breeze seemed to nudge the Princess of God through the last set of doors and thus, help her take her first step on the planet Earth.

At the exact moment Mary stepped onto the Earth, the choir of angels, which had sung so beautiful before, started their songs of praise again. This time they were full of joy, radiant and splendour singing their hearts out, loud and cheerful. Angel musicians joined in the joyous throng of voices, adding depth and volley. The ground, on which this part of Heaven stood, started to shake and tremble with the power of their voices and instruments.

From nowhere, King William V and Queen Kate, in their splendid robes appeared before the church, standing behind First Lady Rocc and her disciples, everyone gasped, even the camera crews turned a brief moment to capture the appearance of the new guests.

As they witnessed Mary before them, they fell to their knees, realising this was a higher order than themselves. They bowed their heads low to the ground, not lifting them until told to do so.

Ladies curtsied low, bowing their heads slightly and First Lady Rocc was the first person to step forward and kneel before Mary, touch, then take hold and kiss the hand of the daughter of God.

As she kissed her hand, a glow encased her body sending a brilliant aurora around her. It lasted a few seconds and was gone. When she stood to face Mary, she smiled, knowing there was no fear in her and nothing at all would hurt her until the day God Almighty decided it was time for her eternal rest.

Archbishop Langley, Bishop Huntley, Bishop Mathews and the other disciples followed next in line, each was enveloped in turn by the aurora and glow. Every breath-taking moment was captured for life on film, passed throughout the world so each woman, man and child could pay testimony to the daughter of God; Mary was truly and finally on Earth, here to stay, until her work was completed.

When the disciples were blessed, Mary walked forward to King William and Queen Kate who still had their heads bowed low. She touched their heads and whispered to them, "Rise!"

They looked at Mary, slowly getting to their feet, aided by Mary's outstretched hands. She spoke to them and blessed them, giving them the gift of multi language speech before allowing them to return to their former duties. The couple vanished from view, just as suddenly as they had appeared, leaving many journalists talking, wondering what was said and why they were summoned here.

Mary talked quietly to First Lady Rocc and her disciples for fifteen minutes before turning to the crowd of cameras and stepping forward to the array of microphones which had been quickly placed in position for her use. She looked stunning to say the least, no woman on Earth looked as beautiful as her, even though she wore the simplest of clothes and shoes.

Reporters, television interviewers and magazine press alike, wished to pose a million questions to her but she stood silent with her disciples beside her as the giant doors behind her closed and the angelic voices faded away, drifting back from whence they came.

All of a sudden it seemed so quiet, even the birds were silent as if waiting, like woman and man, to hear for the first time in two thousand years, the voice of the daughter God.

"Good morning to you all," Mary said softly into the microphones.

Her voice was delicate, soft, petite, yet firm and strong from within. There was no quiver like some expected, talking to the press was no minor task. Facing her public for the first time within a very short period of time of stepping foot upon the Earth was a giant task in itself.

Some women clapped their hands together, tears rolling down their cheeks in full flood. People throughout the crowd and the world were crying for joy as at last Mary was here. It was going to be proved beyond all shadow of doubt, for all those doubting Thomases around the world that had put down the church. The world was about to be told the full truth by the daughter of God. No man or woman dare take her word lightly, this was the moment everyone had waited for.

A short woman, whose face was covered in tears, with black mascara running down her chin, dripping onto her white blouse moved through the crowd of television cameras as if unobserved. Everyone was stood perfectly still, cameras previously lined up and everything on automatic; no one moved or lifted a finger to stop her.

Reaching the front of the crowd, touching the rope, she fell to her knees, she bowed her head low, sobbed loudly and leaned over the thick red and white rope, stopping everyone from passing it.

"Your highness, daughter of God, Mary, I have believed in you since I was old enough to go to church. I have never doubted your word and I knew in my heart you would come to our planet before I died.

"My son, Jonathon, who is 21 years old, lies dying in hospital many miles from here from a dreadful disease called cancer. Please, I beg of you, I am willing to give my life, for which I have lived a long and happy life, in exchange for his. Please Mary, please save my son's life and soul."

She reached out and gently touched the bottom of Mary's robe; she felt her entire body tingle and was expecting to die at any moment. She wished she could see her son for the last time, but knew the exchange of lives was fair, and she had said her final farewell the previous night at his bedside. She knew she had not asked to see him before she went to heaven; it was her loss and her son's gain, a life for a life.

"Susan Evans," Mary said aloud. She looked down upon her crying gaze and held out her hands to lift her up. "Your faith in me is good enough; my being a woman has come as a shock to many people throughout the world. Many faiths have been rocked, torn apart at the very seems, shaken to the very roots of their foundations.

"Your faith in me has not receded, despite the horrendous shock you have all sustained, your son is healed and I herby start the wonders for your planet. Through your faith, your belief in me and my Mother shows in your face and your heart. For you alone, as I speak, every person on this planet, suffering from any form of cancer is now healed. By this time tomorrow, every form of cancer will be banished from the Earth for all eternity.

"Other illness will follow, painful deaths will cease, and for today only, neither My Mother nor my Father Lucifer, will take a soul from this planet. Before you all ask, just like here on Earth, even Gods need a male and female to mate and produce offspring. This is my first of many commands," she held Susan Evans' hands and smiled. "Go to your son, be by his side and share your lives, for they will both be very long and very happy."

Still crying with joy, the mysterious woman disappeared from view, and when she opened her eyes was with her with her son who was smiling and sitting up in his hospital bed. Doctors and nurses were running around not believing what had happened as people on the edge of death suddenly sat up and looked very well. Their bald head's now with a full head of hair and their cancers; of many different kinds, were wiped clean from their body.

"Wow!" exclaimed Jim, the reporter of the Evening Standard. "What a scoop; that was Mrs. Susan Evans, mother of three, and her husband Sam, gets Mary to cure her son Jonathon of cancer. Following that, Mary cures all people of cancer and painful deaths. I wonder if that includes AIDS?" he said to his friend Paul of the Morning Press, his sister paper.

"I guess we'll have to find someone who was or is suffering from AIDS and ask them," Paul replied. Leaving the group of other reporters, the two men hurried off to get their scoop completed.

At the church, Mary was getting ready to start her main talk and answer questions which would be put to her.

***

Twenty miles away; despite the revelations of what God was capable of doing, behind a very strong shield given by Lucifer, more of Mark's recruits were getting ready to spoil the great day. Lucifer, down in the bowels of Hades, was watching his army personal who belonged to the 17th Tank regiment. The ten men had already overpowered their colleagues knocking them unconscious, dragging them away from their six Chieftain Tanks.

They were five kilometres from their barracks and about to start manoeuvres for a peace keeping mission which was soon to start in the Far East. Despite God's intervention they would be needed for helping to clear and build new sections of road, remove barricades and help the Royal Engineers with the construction of bridges, houses, hospitals and schools. Most of all, one of the priority buildings was to be a church. A few months ago, this was considered well down the list, now it was right at the very top, equal with a hospital which the army would build at breakneck speed.

God may have cured some of the seriously ill, but there were still those who had minor injuries and wounds, broken bones, shell shock, associated with other mental problems which occurred in war torn countries. These people would need help as well and some women and children, who had been raped during the wars, would never recover, unless God helped them, but there was still so very much to do.

The men in the group were now each capable of driving a tank and operating it, even firing live shells which at that time, were locked away inside the barracks. Mark had given them super human strength so they could load the heavy shells into the guns of the tanks.

After ensuring one tank wouldn't work, the remaining five tanks started up and headed for their base at flank speed. The short distance was soon covered and the guards at the gate were caught un-aware as the mighty weapons of war trundled without stopping through the metal barriers, crushing them flat, taking the guard's small hut, television included, with them.

The guards managed to jump out of the way with only seconds to spare, they had encountered runaway tanks before but five were unheard off and too suspicious. By the time they had collected their wits, got up from the floor and surveyed the damage, the tanks were at their destination.

The mighty engines stopped and the men alighted from their secure positions with elegance and speed. They had everything worked out; they knew their plans by heart which Mark had worked out for them before his untimely banishment by a higher God than his.

The door to the shell hut, despite it being made of steel and protected by an elaborate alarm and entry code system was wrenched from its hinges by one man, Captain Geoff Lawson, who was the strongest of the team and their leader. The guards on the outside had been knocked unconscious and left in pools of blood. The armed guards inside the building didn't stand a chance as Geoff advanced in leaps and bounds, kicking furiously at two guards knocking weapons from their hands. One guard managed to fire two rounds at Geoff but the bullets simply rebounded from his body, not stopping his forward movement. Sergeant Peter Wilson picked up both weapons and tossed them back along the line and the last man stood guard just inside the broken door to the hut, now with a machine gun in his hand.

The live shells were quickly located and although heavy, were stacked on a trolley and pulled by one man back to the door. Sergeant Adam Hopkins loaded them with Sergeant Dan Green into the tanks as more and more shells arrived. Then they loaded rockets into the tank turret and live machine gun rounds into the tanks armoury. The last things they loaded were small arms and machine guns with extra magazines for them all with a hundred grenades for each tank. They also loaded stun grenades and smoke grenades, enough to cause panic and start a small war. It was their job to ruin the day as Mark had wished. It was their only task in life and they would die carrying out their duty if needed. Lucifer was determined to upset the church and God for what She had done to Him, especially showing Mark up before the world. It was His intention that His wife would pay highly for the cost of His loss.

It took the men ten minutes to load their vehicles to full capacity, food didn't worry them in the slightest, Lucifer had seen to that, they were only hungry for war and destruction. The alarm had been raised and the tanks finally located. Most of the men were away from camp in different locations throughout the world so the few who had been mustered to seal the camp and capture the men, were very small in number, especially as they were going against five tanks now armed with live ammunition.

As the men emerged from the shell house, despite the danger involved with high explosives, the order was given to open fire. No one worried if the renegades died, that was their aim, these people had to be stopped, and the tanks could not leave the barracks.

Bullets flew through the air, ricocheting from walls and steel doors; the renegades retreated for a brief moment inside the concrete and steel building, collecting more weapons and ammunition for their immediate use. While the men inside the building formed a new plan, Peter closed up his tank and started the engine while Dan climbed into the turret and started the turret controls.

The turret turned until it faced a group of soldiers who had inadvertently located themselves in front of petrol tanker. One shell was loaded into the tank's gun barrel and fired; a second later it hit the tanker and exploded, sending burning fuel high into air covering the soldiers who ran around on fire. The tank's machine gun opened fire and put the soldiers out of their pain in a few seconds. Marks' men fought furiously for their right to survive and carry out their duty, nothing would get in their way, their aim was better than the best marksmen the army had.

Soon the gunfire stopped and the renegades ran to their tanks, climbed aboard and locked themselves in. Starting their engines, the tanks moved out of the area and down the road to the main gates where another group of soldiers had already managed to place two Chieftain Tanks in the way off the barriers to block their escape and added two large lorries to help fill the barricade.

The lead tank fired two shells which blew the lories apart and the third tank opened fire on one of the tanks blocking their escape. The stationary tank exploded and a second shell removed the remains of the tank so the five tanks could pass through onto the open road. As the Tanks moved forward, their machine guns took down another ten soldiers who were firing their hand weapons at them. Two men who had a hand held rocket launcher were hiding behind a fence, as soon as they fired the weapon; the crew in fourth Tank fired its machineguns and two rockets which intercepted the rocket heading for the fifth tank and destroyed it. The two men were then riddled with bullets as the tanks rumbled past their dead bodies and rode over the remains of the tanks and lorries.

The tanks swiftly moved from the road to a field, across it through wooden gates, smashing them to pieces, crushing hedges and destroying farm land, turning up pasture and crops alike in the wake of their heavy caterpillar tracks. Now and then they practiced their gunfire on any passing cattle or sheep they could see in their gun sights.

Back at the barracks, the men who had been burned and killed, injured or knocked unconscious were coming back to life, as Mary had ordered, nobody would die on her day, and God kept her word, reinstating their lives, but not until the danger was passed. God could have stopped the tanks, but She wanted to see how the humans would deal with the situation.

By now the alarm had been raised at another camp and a small platoon of men were armed and with helicopters in pursuit, were searching for the tanks and working out their destination. The local police were informed of the impending catastrophe that was heading their way, and were reassured the army were doing their best to get troops to stop the tanks.

Army gunships scoured the area and the countryside in search of the tanks and within ten minutes of them getting to the barracks, the tanks were located heading towards the small town of Radstock.

Nobody connected the coming of Mary with the tank's destination, until one of the police officers, alert and hearing the incoming messages from the army, suddenly realised the potential of what could, or was about to happen.

"Sir," he said rushing to into the Station's Commander's room.

"Constable Perkin, don't you ever learn to knock?" Inspector Andersson bawled at him.

"But, Sir, five tanks are rampaging through the country . . ."

"Not now, can't you see I'm busy, anyway, shouldn't you be checking the crowds and ensuring nothing is happening out of the ordinary?"

"But Sir, the Tanks . . ."

"I will not tell you again, if you continue this interruption, I will have you severely disciplined, even suspended," he shouted raising his voice and tapping his pen on his desk. "Do I make myself clear and fully understood?" he continued, standing up looking very annoyed.

"Inspector Andersson," PC Perkins said, stiffing up his body in an attempt at saluting. "Regardless of your threats, the five tanks which are rampaging through the countryside fully armed with live shells and machine guns are heading straight towards us. They will be on the outskirts of our town in less than thirty minutes.

"May I respectfully remind you that also in our town this morning at Radstock Church, Mary the daughter of God has set foot upon the Earth?" He stood perfectly still, sweat dripping from his forehead.

"PC Perkins . . ." Inspector Andersson paused; and thought over what had just been said and turned to a large wall map of their area behind him.

"PC Perkins, in future, you continue to shout at me if such important news comes through again. Who told you about this? Why wasn't I informed?"

"I just heard the call on the radio and realised it myself, I don't think the army has put both activities together, they are searching for the tanks as we speak. The printer spilled out all this paperwork."

"Alert the army, they must have something to stop them, alert all the officers on duty at the church to prepare for mass panic. Alert the hospitals for a code red for possible incoming injuries from bullets and tank shell explosions."

"Yes Sir," he replied and standing to attention, clipped his heels together. He turned smartly and left the room with a wide grin on his face.

While men and machines moved forward to try and stop the oncoming tanks, a far higher force was waiting patiently for the tanks at the boundary of the town.

Inspector Andersson, PC Perkins and the three other officers draughted in from crowd control, waited in their police cars at the edge of the town in the hope of trying to stop the tanks. Hopefully, they would try and reason with the soldiers while the army got their act together and arrived in force to stop them properly and move the tanks from harm's way.

The three police cars looked pitiful against the mighty tanks as they loomed above them. The huge gun barrels, so small on television, now appeared very large as they pointed towards their small cars with live shells ready to fire and blow them to pieces. Second by second, the mighty machines of war got closer and closer to the police cars which had barricaded the road.

As the men in the police cars prayed to God, a sun beam came down from the sky and covered the five tanks, stopping them in their tracks; no matter how hard the tanks tried, they couldn't move forward or back. The men inside tried firing their shells, but none worked. Then they tried the large heavy machine guns and they refused to fire as well, as too the rockets refused to leave their launch tubes. The soldiers inside the tanks were at a loss as to what had gone wrong. They called to Mark to assist them, but he was not available, not in the location of Earth anyway.

A squadron of heavily armed helicopters closed on the tanks position with army jeeps carrying a Captain and General getting closer by road. A battalion of men, who were swiftly mustered from another camp were following in trucks, racing along the roads to try and intercept the tanks.

As the helicopters closed on their targets, they prepared to fire their live rockets. More sunbeams, lighter than the one over the tanks, covered each helicopter, as the pilots fired their anti tank rockets, nothing happened, once again machine guns also refused to fire no matter what the crew did to try and make it happen. The helicopters landed behind the tanks and the army personnel inside raced towards to the tanks, their tracks turning but going nowhere, the lush, green grass being churned up beneath their heavy, metal tracks.

As the soldiers raced towards the tanks, hand guns ready to fire, the drivers of the tanks turned the engines off and climbed out of the machines and ten confused, frightened men, stood in the field. They looked around, wondering how they had got there and trembling with fright, were told what had happened and the men they had killed and then God had brought them back to life.

When the General Briggs arrived, he was smiling, and looked very pleased with himself as he told the men they were not accountable for their actions as they were under the control of Mark, and their sins had been forgiven by God. If God could forgive their sins, then so too could the army. The soldiers looked much calmer after being given the news they had not killed anyone and that not one fence or animal had suffered while they were under the influence of Mark.

Captain Geoff Lawson stood to attention when he faced General Briggs for the final time. "Captain Lawson reporting for duty Sir; it was not our fault as I am sure you now understand, under the circumstances, we will do whatever you say Sir, whatever punishment you deem to give us, we will accept without argument."

"You must have more faith in God, but I can see in your eyes, you are now changed men and all of you have been touched by God's hand. There is still that posting abroad . . . if you and your men will accept it?" General Briggs said smiling.

"Of course Sir, we will go wherever you send us; however, instead of taking tanks, we would like to take lorries and help the people, rather than bring more weapons of mass destruction to their countries. Sir, there is one other question I would like to ask."

"What is it?"

"You said you could see in our eyes that we were changed men, what did you mean?"

"Ah, you have not realised, everyone who has been touched by God, has a brand new colour eyes, like mine, yours are now electric blue." He leaned forward with his new forty, forty vision and showed him the colour of his eyes.

"Are mine the same colour, Sir?" he asked.

"You have all been touched by God, as was I on my way here when God explained what had happened to you and what she had done to stop you getting into further trouble."

Captain Lawson turned to look at each of his men and noticed the colour of their eyes and smiled. Turning he faced General Briggs again.

"Captain Lawson, you will be put in charge of the entire operation, take with you what you think you will need and any further equipment will be yours as soon as we can get it to you. I will inform the Battalion Commander the change of the operation. We will now call this, 'Operation Rekindle', where we will go into Africa and join people together and build them homes in which to live in peace and schools to educate their children and hospitals to heal their sick. I am sure the countries around the world will chip in and help with materials and finances.

"If there is one team that can accomplish this operation, it will be yours, and it will be completed without further bloodshed, accidents accepted of course. There is an air of gratefulness about you now that was definitely not there before this episode. I must say, it is a change for the better, with God now taking a firm interest in our world affairs, I am sure the forces and its equipment can be put to far better use than the weapons of war we have at the present," General Briggs added.

At that moment, his body was surrounded by a yellow golden light which swirled like a whirlwind about him. As the others moved back, they could see he was looking at himself, obviously not being harmed in any way, and by the way he was acting, he was listening to the voice of God. A minute later the glow disappeared and he was left smiling, almost knowing the future.

"Are you alright Sir?" Captain Lawson asked stepping forward to check his General was okay.

"I've never felt better in my entire life thank you."

"Was . . . God speaking to you?" Lieutenant Adrian Banner asked with trepidation.

"She most certainly was; a very pleasant voice it is too. I am to go to London and report to the Chief of Staff; there I am to advice him of God's plans of what She wishes us to do about the armies, navy and air forces of the world."

"That is tremendous news Sir; a great honour and responsibility has been placed on your shoulders by the highest authority there is; I am very pleased for you, Sir," Captain Lawson said smiling.

"Not only that, but I was told other generals have been told the same as myself all around the world, God is helping us to change the way in which we live, and we have to do as She wishes, and I know we will do it. We will make this a far better world in which to live.

"Captain Lawson, you are now to be a Lieutenant Colonel with immediate effect, congratulations." They shook hands and both men smiled at each other.

"Well done, you may speak to me whenever you like and request what you need," he said using telepathy.

"Is this two way?" Lieutenant Colonel Lawson asked.

"I certainly hope so, or it won't work and once you are better acquainted with this form of communication, you will find other officers have been given this gift around the world so that contact is much easier. You will also be able to speak the native language of the people you meet in their country; that is what I have been told by God just now. Good luck and God speed," he added using telepathy.

"Prepare your troops and your equipment, we'll leave the tanks here, I'll place a full squadron of heavy air force and army air transporters at your disposal, take whatever you need and the right men for the job. I also suggest you take a few hundred civilian builders with you and I have also been told that a number of women designers have already or are right at this moment, working on designs for portable buildings to have built here and transported to you within a week. You will leave in forty eight hours and ensure you have a field hospital and two field canteens with you. I will ensure you get all the food you need from various countries around the world to assist you in your work. God help you speed your mission forward and be with you at all times, as will I." They shook hands again and touching his head with his finger, winked at Lieutenant Colonel Lawson.

As his last order in the field, General Brigg told the soldiers standing behind the tanks to guard them carefully until the army transporters got there to take them back to the barracks by road. Looking back across the fields, the torn up grass and demolished hedges were now back to as they were, only the hedges were now much greener and so too the grass, in fact, the entire field was much greener than it had been in years.

"Inspector Andersson," Lieutenant Colonel Lawson said. "Would you mind leaving a few men here to stop any children trying climb on the tanks, I know the soldiers are here, but a police officer always has a lot more power with children and they will listen to your men I am sure. It will be a few hours before the transport arrives to load the tanks; I really am sorry for all the inconvenience and taking you and your men away from the day's special event. However, I don't think Lucifer or Mark has finished yet, you had better keep your eyes and ears open."

"Of course, as you put it that way," Inspector Andersson agreed, slightly confused and glad a catastrophe had been averted. The thought of what those tanks could have done to all the people at Radstock Church was beginning to haunt him and make him shudder.

Watching the men drive off in a lorry which had just arrived, he returned to his officers and gave them their orders before returning to his station to make out a full report.

Deep down in Hades, Mark, who was still recovering from God's punishment, watched as one after another his plans to disrupt the proceedings failed in dismal horror before his eyes. Even his Father was becoming annoyed with him for not taking more elaborate steps to ensure the day was a total disaster.

***

At Radstock Church, Mary was now in full control of her senses and the police had removed ten more people who were Mark's dark angels. As soon as Mark's agents were discovered and removed to a quiet location, Nigel stepped in and restored their soul to as they were before. Each time this happened, the person was changed to a more loving and caring person and in some cases, given the gift of multi-language speech, to assist others in their town and were given a new role to fulfil their destiny.

The press were now well acquainted with Mary and the stiffness was gone from everyone. She was now seated with her disciples surrounding her before the closed doors of Radstock Church, now the most famous church in the world.

CHAPTER 14

The last adjustments were made, the lights corrected and the decisions made as to who would do the interview. As one man originally believed in the story and got things moving, it was decided that Mike Adams would do the main interview with Mary. It was after all what he had started, so it was his job to see it through to the end.

"Mary, now you are here on Earth, what are your immediate plans?" Mike asked.

"To meet people, let them see me in the flesh, so to speak," she chuckled. Even the daughter of God had a sense of humour.

"Throughout our planet, we use money to buy things, as you have no real job as per say, where you will get money from to purchase things? No one will deny your job is long and hard, we know what Jesus, your brother went through, but what I was trying to get at. . ."

"What will I use for money to pay for things like food and transport?"

" . . Yes. . . I do," he said softly, feeling embarrassed and paused briefly to hide his reddening face.

"Since my brother Jesus was on the Earth, the world has got a lot bigger in terms of size and population. Technology has increased to a state where you can travel to the Moon or Mars and now out past Pluto and the enclosure of this planetary system. You have orbiting satellites, even the screens My Mother supplied were never dreamed of in my brother's day. However in Heaven they were available, we have all forms of technology you have yet to discover. How do you think you made all these fantastic technological advances?"

"Some people are extremely bright and invent things?" Mike replied, hoping he had answered the question correctly.

"Correct, but the idea has to be there in their mind, the ideas come from what the soul sees in Heaven and what God has ordained the person should take with her or him when they arrive on Earth. We do all forms of research for those interested in the subject; however, it is a mixture of us and Lucifer which makes weapons of war. Turning the good we make into bad; like your atomic bombs, which by the way, have all been defused."

"What!" exclaimed Mike amazed.

"I shall deal with the military and all their tasks in due time, some military people have already been spoken to by my Mother and these people are already changing the way the forces think and work. While I am here, God has disarmed all nuclear weapons, cleaned up the atmosphere and nuclear power stations. You still need power and she intends for you have it, but the power stations are so much cleaner and no longer pollute the air we breathe. Also, you no longer need as many power stations as God has changed the way nuclear fusion works and has given your nuclear engineers Cold Fusion, you can now run a hundred of your cites for free, and people will no longer be forced to pay for their gas, oil, electricity or water, all these fuel sources will be free from this day forward. God will see to it that nothing goes wrong and all the pipes and cables are up to date and will last for many thousands of years."

"That is a miracle in itself," Mike agreed. "We have been trying to get the Generating Boards to do this job for years, but they said it was impossible and the costs far too high."

"The costs were never high, they had all the technology they needed many years ago, they just didn't wish to shut down the power stations and make the modifications to lose money. The boards of these companies are now fully aware of the changes God has made and are right now getting the letters organised to tell everyone they no longer need pay the power stations and companies for their power, gas and water. These services will still need people to do work, but it will all be paid for in a different way."

Mike paused, he was not expecting this and the question of finance had somehow been overlooked. He knew as his producer was screaming into his ear, it had to be returned to.

"Mary . . . going back to finances, how do you intend to finance your stay here on Earth? I refer to hotel rooms, travel and I presume you will need to sleep?"

"I do not need to sleep, but I will do so at times to blend in. I have funds believe me, as to material objects, what would I do with them? My time here will be spent with my Mother's children, my sisters and brothers, as for food, are any of you hungry or thirsty?"

"No, but we have eaten and are full." Mike looked at his watch; it was now just past eleven thirty, coming up to lunch time. He thought deeply, he wasn't hungry, not even thirsty, yet the last time he had eaten anything was five that morning and at that was the last time he had a drink as well. He knew he should be getting hungry or at least thirsty.

"I take it you are responsible for the food inside us at the moment?"

"Of course, the food of life; let me put out of your misery and stop John from shouting into ear. There is no need to shout John, I can hear you perfectly well," she said to the producer a hundred metres away in the VT Tent.

"I'll answer your questions directly, I have at my disposal enough money for myself and my disciples, the church is my home and God has built a special home for us in each county of the world. You cannot miss it, it's pure white and very large. It has a large staff and my people are there to answer people's questions with whatever is troubling them, including health and money problems. If they have trouble understanding what is happening, that will be explained as well."

"Where is your house in this country?" Mike asked.

"It stands near Silbury Hill, there on the large screen is one of my houses," she said pointing to the screen which had appeared to the right of Mike and was facing him.

It was indeed a very large house and pure white; the grounds were extensive with trees and flowers, cars and people, animals of all types running free. Birds of every type were in the trees, many nesting and in the large lakes; swans and ducks were swimming around and making a triumphant noise.

"I have never noticed that house there before, and the land looks so large, at least twenty acres at least."

"It is exactly ten square miles, it was put there a few minutes ago; nothing was there before except a few derelict buildings and land which has not been used in eons. This small town is covered with derelict houses and land; besides building this house, God has constructed houses of one, two, three and four bedrooms on every piece of spare land. Shopping centres and factories with equipment to make clothes and material objects that people need are ready to occupy."

"Who will occupy these houses and factories?"

The people who will be first to take the homes will be the homeless and those in great need. As priorities arise for the three and four bedroom homes, so they will be filled. Also, every house in this town and five local towns have all been renovated to the highest standards and the gardens now grow plants and trees to help keep the air pure and allow the birds and squirrels a place to live. All these homes will now be rent free, they will not need to be repaired, God will see to it. A small council force of people will now ensure each house is filled with the right people and there will be no need for anti-social behaviour rules anymore as all children and adults alike will obey the word of God. Neither will there be the need for prisons, as nobody will need to break the law, they will not wish too or face God's wrath. Only God will know what punishment if any She will give the person, but the police will be there, in small numbers to assist in accidents and help people in general. Just because I am here and made a few rules for today only, will not stop people from dying or getting injured. We will still need hospitals and staff to help the injured people, but the main diseases will be banished from Earth and as John thought of the question several hours ago, it does indeed include AIDS.

"Returning to the homes and factories, all rent will be free, there will be no rates to pay either; the councils will find they have enough funds to pay their employees. The houses are being filled as we speak and people are being moved from what you call hotels and bed and board and hostels to the new houses. It is God's intentions that in this country and all other major countries of the world, there will be no more homeless people and people will no longer wish to be homeless."

"But surely, that is the way our society has been built?"

"Then it will change as of today; society will change, people will change, people will become equal in every way. There will no longer be the very rich and wealthy and the very poor and penniless, not in these advanced countries. Wealth will be spread evenly throughout the land; all mortgages are now cancelled, whoever owns a home, does so from this moment forward. All Building Societies and Banks have now handed over the deeds of the properties to those people who live in them. People will no longer need as much money to live on; the wages will go down, because people will not need so much; where problems occur, I will step in and solve them, or one of my disciples will do this," Mary explained.

Mary was showing Mike that God had had enough of Her children being pushed around from pillar to post, made to feel inferior, having to get loans to pay for the very house they lived in and need to keep from getting wet and cold and to grow their family. The government would no longer push these people around, the government and King and Queen would change their ideas and make new laws and banish old ones," Mary explained.

"Some people will say this is unjust, unfair to those who have either worked hard all their lives or been born into money and have a right to be in a higher society," Mike replied.

"You own your own house don't you?"

"Yes I do," Mike replied.

"You are also six months in arrears in your mortgage repayments; up to your eyes in debt and your wife has to go out to work to help feed your children."

Mike blushed profusely and looked around at the others staring at him in amazement as he tried to hide his face from the camera. "That is correct," he whispered.

"You are no longer in debt, you own your house outright, and all your bills are paid. The money has actually been paid into the building societies and you have the title deeds in your hands right now," she said smiling, almost laughing and looked down at his hands.

Mike looked down and sure enough, as he unfolded the document in his hands, there were the title deeds to his house. Letters from banks and other creditors informing him he owned them nothing.

"But how?"

"This day everyone will start anew; your monetary system is old and well out of date; like you say, it's unfair of God to furnish the old and homeless in new accommodation, so to be fair, all people all from this day, own their own home.

Councils will have new homes; everything will become clear as the day progresses; the disabled will walk, see and hear; everyone will have a new lease on life, a second chance. Some of you will be designated as informers for God, they will give God's plans to the people God wishes and they will be produced and sent out into the world. You are now living on God's world, it is not yet perfect by any means, but She plans, with your help, to make it that way, to push aside evil and Lucifer forever."

"Does this mean . . . Does this mean that Lucifer will no longer exist?" Mike asked.

"No, what it means is, with your help, we will drive Lucifer from the bowels of Hades to good, bring Him unto the Light. I want Him to see what mistakes were made and perhaps, with His aid, stop the world from falling into the chaos it was heading for," Mary said softly, really hoping Her words would be heard by everyone, including her Father.

"Without darkness, what will man and woman have to be here for, what will our purpose be in life?" Mike asked Mary.

"To live of course, exist and be happy; to love each other and prove, in a material world, with material things, love and happiness, together with belief, can coincide until the day of Judgement. This will be when God Herself comes to Earth to judge the good from evil for the very last time.

"For the moment I am here to do whatever I can to help people change their ways; assist their needs and bring about the change which is needed to make this world a greater place in which to live. There is a whole universe out there to explore, and explore it you will, for its God's wish you should expand into space. There are many new civilisations to greet, to spread the word of God; there are some dimensions that only a few of you have ever dreamed of to visit and wonder how you will get there. To pass through Black Holes into a parallel Universe which coincides with this one," Mary said making it all sound very easy to accomplish.

"That all sounds very exciting, but we are only just getting into space. We have perhaps two or three centuries before we can get to the stars and design light drives with our current space scientists. Our current rockets only just get a few tonnes of material into space, let alone millions of tonnes of material to build huge spaceships."

"Actually, it is much closer than you think, much closer than two hundred years; you are on the verge of cracking interstellar flight. You will very soon have ships that travel faster than the speed of light and soon you will travel through Black Holes that fold space, joining galaxies together. Many of the new factories behind us on the big screen will be making parts for these massive ships which will be over three kilometres long and have a crew of several thousand and carry thousands of scientists and engineers and most all, my disciples through space to new worlds. As I am speaking to you, women designers are creating the drawings and engines designs for these very special ships.

"Space engineers, who have wondered on these problems for years, have now been given the equations to work out what they need to know to power these ships. They have been given a helping hand to speed these things up. Some of my disciples sat here this morning are waiting to go on the first ships to the stars and visit new planets; they will be among the first people to see new forms of life which breathe methane gas, not oxygen."

"What about Lucifer . . . . Is He likely to be on other planets, to assist in their final destruction?"

"Unfortunately, yes, like myself and God, Lucifer travels the depths of your space, even into other dimensions which are hard to explain unless you understand the concepts of space and some of your people do, you just don't listen to them, despite their attempts to try and tell you all about other dimensions. These people will also be among the first ones to go deep into space and reap the rewards of their trials. You have been given life to explore this universe, or multiverse, to visit the planets and leave this solar system and travel in God's creation."

"Then how close is space travel?" Mike asked excitedly, trying to discover more, even though there were still a million questions to ask.

"The first trip to the stars is less than two years away and the ship will be three kilometres long, one point five kilometres wide and a kilometre and half high. It will be built here on Earth and fly into space using antigravity engines. It will carry the first disciples into space and I will give your people the location of many planets they can visit close to Earth which are inhabited."

"It was too much of shock, Mike sat back in his chair, he was mesmerised by her words. Man had barely gone to the Moon and was now on Mars, although astronomy was now a subject in the school curriculum, there were no trained astronauts for such a trip. The next manned trip to Mars was in seven months, to take the number of people there to fifty for MARS ONE. Yet here was Mary saying the first starship was less than two years away.

"As a matter of interest, how many people will be aboard the first starship?"

Mary looked deep into his eyes; she could see the desire for a reporter to go on this trip. The sudden need to record history to be done by a professional team, if possible his team, Mike dearly wanted to travel into deep space.

"There will be a ship's crew of eight thousand, including shuttle and small ship pilots, ten thousand women from all countries, five thousand men from all countries, plus disciples and scientists and a new job for people, Trade Negotiators. There will be at least two hundred disciples, some from my group here and four Captains to the ship, and each captain will be responsible for a different section. There will be a large number of shuttles and speeders, small one or two person ships to leave the large ship and explore small spaces and do other forms of work."

Mike was stunned and silent, as too John his producer, who up until now had been pushing him as hard as possible to get in question after question.

"When?" Mike asked in a whisper.

"Twenty months away, just twenty short months, now to the other questions."

Mike remained in shock for a few more seconds, then John shouted into his earphone. "Wake up Mike, ask the next question!"

"Sorry, yes where was I?"

"On the subject of what I am doing here," Mary said with a smile.

"Yes, quite, will you elaborate on that?"

"Well, I am here to help and guide you all in your beliefs, to bring them all together and show where you have gone wrong. Not that there is any belief which is wrong, but they are not all right."

"Wrong, does this including Christianity?"

"Christianity itself, like all other religions was not meant to be prejudice as so many of your vicars and ministers are; it was not meant to be so strict. World population has to be curbed to allow those alive to eat and survive. There are so many people in this world, who starve, yet there is enough food to feed everyone, and we return to the point of money once again. These poor countries cannot survive because they cannot afford the food, they cannot afford the technology which many countries have to feed themselves.

"Christianity was meant to interlink with people, because someone is black, does not mean they do not believe in God. It is the belief that counts; the belief in God, any God, for God goes by a multitude of names. It does not matter how many times a day you pray to God, it is that you pray and what you say and believe.

"It does not matter if you eat fish or even work on the Sabbath, have you ever thought about the priests and vicars who have to work on the Sabbath? The nurses who care for the sick? The police, fire and ambulance services who work on the Sabbath? The power workers and people who cook the bread you eat? Those who work in shops where you buy food, the food you forgot to buy the previous day? Your wife; she works continually, more hours a day than many men, Sabbath or not.

"As long as you pray to God, is all that really matters, material objects seem to have overtaken God in the western world, supposedly civilised countries and even come before Her. She is forgotten among many Christians, only remembered at Christmas, then many care about the money they have to spend on other people. They wish to spend it on themselves for materials they don't really need.

"I only managed to get here because there was a sudden upsurge in the faith of God, because churches throughout the world were expecting the second coming at this time. More people started to believe in God so I was allowed with your help; especially you Mike to come here and help push Lucifer back to where He belongs. That is until he repents and becomes the God He once used to be many millions of years ago," Mary explained hoping everyone could understand her.

"We are all pleased you are here, will you be speaking to the leaders of the church for long?" Mike asked.

"Long enough for them to understand where they went wrong and how to change and I will assist them with the change."

"What about other faiths and beliefs like Catholic, Buddhism, and Gods in the far eastern countries?"

"Many of these beliefs are combined into one and they do indeed believe in God, they worship Her in another way. To you it is strange; you call it un-Christian-like, putting it down. The main thing the Pope got wrong was the position of contraception, the bible was interpreted incorrectly; it was a mistake on their behalf, even Christians who put down second and third marriages, a thing of which has become normal today, and is used a great deal. Sometimes, it is better to be loved and lost than never loved at all. Just because you divorce, does not mean you are displeasing God. When the bible was written, there were not the stresses and strains on a marriage as there are today."

"You're right there, even in my own job, we are always under a great deal of stress," Mike agreed, making a point for the people at home.

"Quite so, in the days when the bible was written, the population was small, there were no big decisions to make, time was not critical. There were not many material objects like those of today, no cars, large houses to purchase, televisions, in days of Jesus, life was very simple. If you lived in those days you would find life boring; you wouldn't cope with the peace and quiet. You were not threatened with violent wars of such devastation like today; times have changed, so too has the vision of God.

"God has come to realise in this life; women are becoming equal to men; like they should be. They now get out and see more of the world, carry more responsibility and now, men are expected to help around the house and with their children. In days of old, this would never have happened; yet today, it is expected of men. As men and women have changed, so too should the bible, but it was not re-written and the old word was adhered to. Despite the way in which the world adapted to new life, the bible was also meant to adapt to the new way of life, new times, as women and man progressed through time."

"Then it is not indirectly our fault of the ways we currently pray and interpret the bible?"

"No, that is why God has decided to make Herself known to the world through Me. The bible as you know has been rewritten by God; not a single old style bible remains in the world except one, in a glass case which is impregnable and can be read by anyone. However, there is a warning that this is the old style bible and not true with the current way of living. It is a false book which should have changed with the progression of time."

"Where is this book?" Mike asked for the entire world.

"In a museum created by God for the preservation of old times and relics. It is in Israel, which is to be the New Capital of the world."

People around them, especially the church leaders and minsters listening to Mary's speech were aghast that God had chosen such a place to create the new capital of the world. War was constantly fought here through land rights. No matter how the world tried to intervene, fascists continued to fight for that sacred land which was given to the Israelites many thousands of years ago. It was now to be the centre of attention once again, now it really did have something worth fighting for.

"In this new time of peace, the capital and the main city of this world will in future be known as CONCORD."

"That is an honour for the Israelites; does God have a reason for picking this country?" Mike asked, wondering deep inside if there could be an answer.

Mary paused for a long moment. "The reason behind this, which I have just been told by my Mother, is quite simple. In the early days of Earth, the first civilised people lived in and around this area. For a great many years, it was the only place on Earth which was habitable by humans. As the years passed, man was placed throughout the globe, but it took time for the northern and southern continents to become warmer and support life of any form. It is where the Old Testament was first written, you never hear of England or America being mentioned in the bible, yet man assumes life existed in these countries . . ."

"Or else there was a mass exodus of people to the west, south and north at around the same time. But of course, there would have been the vast oceans to cross and no one had any idea about what was on the other side, if indeed there was another side," Mike interrupted, partly answering his own question.

"That is correct Mike; you have foreseen the true answer. It was therefore the most logical place in the world for the capital to be."

"But it's a continual war zone! Land is fought over day and night, people there carry guns all the time and fear their enemy," Mike explained to Mary, as if she was not aware of this.

"Not any longer, this part of the world will be at peace for all eternity; God has created the boundaries for this land, and there is nothing to fight over. I can show you what this new capital looks like on the screen."

Everyone, even the television crews turned to face the screen. The view was just like that of an everyday modern metropolis in any western country. There was a large runway for aircraft, roads leading off to different destinations; hotels, shopping malls, offices and the interior of the airport security systems was shown next.

No troops, only female guides, dressed in modern western styles; what the rest of the world had come to live with and expect over the past twenty years; but there was no security; no police with machine guns as there were in western airports.

"Everything is automatic; you cannot get into the system because there is no system to break into. As the plane lands, it flies onto a runway which turns drugs to sugar, weapons to toys, bullets of kind to sugar mice. Explosives of any form to dust and grenades to jelly, the mind of the person is checked and if he or she has any ill feelings towards anyone, they are cleansed of these feelings and changed by God. Liquor is allowed in small quantities but not vast amounts or else the entire haul is turned to water.

"Cigarettes are disintegrated even if being smoked at the time. The smoker will be cured instantly of smoking, drinking and taking any form of drug, medicines are allowed of course and any ailments the person has will be healed. You will always need your practitioner's hospitals etc., but that is for the future. Only ailments of today are being cured. Some will be wiped out completely but there will always be minor ailments, which you need to help your immune system when you visit other planets keep you fit and well."

"What about children's ailments? Some of the real bad ones parents fear the most?"

"All serious ailments will be gone forever, Polio, Whooping Cough, Croup, Measles all of kinds gone from the face of the Earth from today," Mary replied.

"The city looks very good up to now, just what we would build ourselves," Mike said turning back to face Mary.

He was now very intrigued with what Mary was saying, this new world was much better than the war torn planet he had lived on for so long. For a brief moment he recalled the times he had put his own life in danger when capturing news reports in a battle zone. In the front line firing areas here, even women and children were killed without mercy.

Mary smiled, looked about her and took a deep breath; the air was so clean and fresh, the people were sat quietly, listening to her talk, like in the days of old as her brother talked two thousand years ago. She smiled to herself, recalling the times; she had watched and listened with great interest to Jesus as he told his favourite stories on a hill to thousands of people.

"The museum," She continued grabbing everyone's attention. "This will be free to enter as is everything in this city. Here you pay for nothing, because nothing can be gained by its control. In future, people who need help, organisations which give aid to vast areas of the world, will be centred from here; where food, heat a roof over your head is all free, even transport is free."

"This is some city, it puts other cities to shame, despite all our technology," Mike agreed.

"Yes it does, but in time, things will change; your planet will change, as too the people with it."

"But companies sending people here will expect a lot of money for flights in and out. They will say they have to arrange everything for people."

"This will be arranged through my people throughout the world in small offices. Planes will be free, as too the arrangements, the number for these companies is 888000, it is now embedded into everyone's head, even children will remember this telephone number and the number is of course free from anywhere in the world and has five million lines," she added with a warming smile.

"What if the telephone companies throughout the world cut you off?"

"They cannot do that; even phones that have been vandalised or have no battery will still get you through. Even if you cannot see the display, just press the numbers and the phone will ring." Mike suddenly held a phone in his hand, it was a smashed up mobile phone with no battery.

"Please Mike, press the numbers," Mary asked.

He did as he was told and a woman answered him explaining about the trip he could take to

Concord. He handed the phone to his sound man who fed the voice to everyone who was listening throughout the world.

"I see, you certainly have some organisation, I didn't think that many people alive were capable of operating a system like this," he chuckled.

"I never said there were, there are more people dead than alive, there always will be and they wanted to help."

Mike was stunned by her answer. "Then, the people operating this system are all . . ."

"Passed on, is what we like to call them, passed on to a better place; a place where there is no fighting, starving or pain; a place filled with music, fine arts people always wished they had time to do. Unfortunately they were always too busy making money, or earning money to pay bills to have the time to enjoy their lives."

"I see, if Heaven is so good, why doesn't God allow us all in there now?" Mike asked.

Mary paused again, as if getting a message from high above.

"You have to live in this universe to appreciate what you haven't got time to do while you're here."

"I see," Mike paused for a moment, getting a message through his earphone.

"Returning to women and men, may I ask how we should have known that God is a female, about women being the prime species on Earth?"

"There were many clues given to you, especially the men that explained things were not as they should be if God was truly a man, the so called dominant species."

"Would you care to elaborate on that?" Mike asked boldly, having no fear of Mary now.

"First, scientists now understand that when a child is created in a woman's womb, the baby is female for most of the nine months, it is only in the later part of the baby's growth, that the sex is determined; until then it is female. That is men have some female organs, but in the male, they do not form.

"Second, in puberty, the girl develops faster and is more 'grown up' about things. This can be by as much as two or three years before the boy, who never really catches up. He therefore uses his masculine strength and ability to prove himself as a man and be better than a woman.

"Third, it is only the female which can reproduce, it was therefore obvious the female was more complex and needed more intelligence to bring up children.

"Forth, this is not through work habits, because a woman in the house does more work, works longer hours and has very little rest or reward for her duties than a man. Therefore it was ordained that women should live longer than men, it was intended in her latter years, without a man, she would have time for her own personal leisure; a time to do as she wished, not what others wished for her.

"Women can cope with trauma far more confidently then men; in divorce, it is the woman who rarely goes to pieces while the man is lost. He knows very little about self preservation, when it is inbred in the female to survive among all things. The female can speak to people where, especially in this country, men tend to be shy and inverted. They are too shy to express their feelings; they rarely kiss other men and consider putting an arm around another man as being gay, whereas women do it all the time. They even kiss each other, showing expressions of love for each other in a different form of love. Men lack the expression of greeting people and sharing affection, in this country especially. It is their downfall and accounts for the higher rate of divorce."

"Really, I never knew; nobody could have dreamt this would be possible in this day and age," Mike replied, showing definite shock. "I do apologise for interrupting you, but you must expect this of me being a reporter," he added smiling, hoping she was not offended.

"Of course, men think crying is for girls, yet it is a form of self preservation; a way your body copes internally. Crying helps the body to survive the trauma it's going through, it helps by reducing your blood pressure and stops you having a heart attack.

"Last, and by no means the least, only a few of you here have been told the way your soul operates and now I am allowed to tell the world," Mary said and glanced at her disciples.

"In brief," Mary started. "When the infant is just about to be born, passing down the vaginal canal from the womb to Earth life; in those last few seconds, the soul enters the baby's body. It leaves through a tunnel going the other way to the death tunnel which is the connection between Heaven and Earth. It is the only way in and out of Heaven; if the soul doesn't make it in time, the infant dies. Without the soul, the body has no personality, with no soul to drive the body, it cannot survive."

"Wow!" exclaimed Mike and there was an expression of wonder throughout the crowd.

"As for the children born in hospital by caesarean section, we know what will happen and the soul jumps into the body just before it is lifted from the womb. If it has been waiting for too long, the soul gets agitated at being confined and causes the mother distress and the baby too; this is what brings the doctor to make an emergency caesarean operation and get the baby out before it too dies," Mary said making the truth known to everyone.

"This is amazing, something we never knew; so man has indeed missed many vital points which should have told us woman was indeed the superior race. Had he realised this, then he would have known God was indeed female," Mike replied.

"Yes, I agree, there were some men who guessed this, but all had their theories put down, covered up many years ago. It was not man's fault; Lucifer had a great deal to do with it when He took the Earth for His own and ruled it as He saw fit," Mary explained.

"Can you tell us what relationship, if any of course there is between your Mother and Lucifer? You said earlier He is your father, so what actually happened, if you are allowed to tell us?"

Mary paused for a considerable time, as if in a trance, looking through everyone, into whatever lay beyond the great unknown.

"My Mother has explained as she and I have exposed many people here today, it is only right you should know the truth as well. This is the only way forward, for you to believe in God, God has to be truthful with you all.

Mike smiled and said nothing, but John prompted him as hard as possible through his earphone. Mike still remained silent; he had to give Mary the time she needed, her moment to explain the intricate details of a life gone by millions of years ago. For this moment, a few more seconds would not matter.

"Many millions of years, long before this planet, this universe was even considered, they were both very happy . . ." She paused and looked up to Heaven.

The sky was suddenly black, blacker than coal and the light from the sun, no matter how bright it shone, could not break through to penetrate the blackness which was forming between it and the new child below. Inside the black, bright blue streaks of high voltage lightning was flashing to and fro. Thunder; louder than any heard over the planet before, was erupting inside the darkness.

Powerful stage lights were switched on but the blackness absorbed their light; it looked as though the show was about to end here and now, all because of an open wound of long, long ago, that had not been healed or talked about in so many eons.

"Who is doing this?" Mike asked Mary urgently, already knowing the answer. He was afraid, he had never been in a situation like this before and a voice was shrieking into his ear.

"Ask her to stop it," John roared above the deafening claps of thunder.

A wind blew up, threatening to disrupt the events even more, but not a flower or petal moved out of place, not even a leaf dropped from the trees and the birds remained exactly where they were, despite their very light weight.

The huge doors of the church slowly opened behind them and a brilliant light shone forth, lighting the entire area. The light diminished to a brilliance the cameras could cope with and everyone's eyes could handle and Mary was told to continue. As she started to talk, three more gold crosses moved on their own out of the inner chambers of the church, floating steadily over the path. One moved to the east, another to the west while the third moved south of the very large gold cross that was now hovering over Mary. Mike and the camera crews were far too interested in what was happening above them rather than what Mary was trying to say. She had to snap her fingers to bring Mike's eyes back to her.

"There is nothing to fear, God will protect us, Lucifer is annoyed and extremely angry at what is being brought up. He doesn't like to admit, that once, eons ago, He and Mother were husband and wife and in reality, they still are, as they have never divorced, but are separated by a fine line that separates Heaven and Hades. Lucifer is my Father, as too He is the Father of Jesus and Mark."

An enormous clap of thunder sounded over the church as lightning struck brilliantly between the clouds. The four crosses turned on their backs and radiated their love upwards to the black clouds. In a matter of one minute, everyone witnessed the clouds shrink in size; the lightning stopped and the sun shone again, lighting the Earth as it had since God placed it there.

"I take it Lucifer doesn't agree with you talking to the likes of us about His past?"

"Correct, He does not like the truth to be known about Him, that He become jealous of Mother's love and attention for your people. Her talent for creating things of beauty, so when She created the heavens, Earth, other planets and their life, stars and women, Father took over. He forced man to steal everything which was rightfully woman's in the first instance."

"Why didn't they just divorce like we do on Earth?" Mike asked.

"A God's marriage cannot be broken, they will be forever husband and wife; I am sure deep down inside, He still loves Her. Like many of you, He does not like to show His affection, I think this is one of His traits that He passed onto man."

"When a man gives a woman flowers, his friends make him feel silly; especially if he has to walk home with a bunch of flowers in his hand. Did you know; men send flowers through shops rather than run the risk of showing themselves up with their so called mates, even if it happens to be her birthday or their wedding anniversary?" She looked at Mike seeing in his thoughts the number of times he had sent his wife flowers through the same problem.

"I didn't realise it, no; but it doesn't surprise me, my friends actually do the same to me when I take my wife flowers. I suppose it's something in a man that makes him feel like a woman," he admitted.

"There is part of a woman in every man; after all, he started out as a woman. Some men are more female than others, some take the ridicule and laugh it off, thinking more of their wife which is as it should be."

"Are there more than just two Gods not including you, Jesus and Mark?" Mike asked interrupting again.

He was amazed he asked the question, he was sure it wasn't from John and he was sure it wasn't in his own head. In fact, he wasn't sure if he actually asked the question. Mary gazed deep into his eyes, they were normal now but a few moments ago they were black she was sure. She paused before continuing, getting advice from her Mother before speaking.

"Since it was not indirectly you who asked the question, I have been told it was not a proper question. This form of question would never have entered your head. Therefore Father, I have been told inform you to ask this question yourself at a later date. You know the rules; you should not involve mortals in the work of higher authorities."

Mary didn't look directly at Mike, she looked through and inside him, deep inside to where a beast of long ago, was scurrying back to the darkness of His home. The answer from Mary echoed through a cast of chambers, continuing down through the darkness until it reached the bottom cavern where Lucifer lived. The message echoed all around Him, hurting Him not to know the answer, to be able to discover the truth.

Without warning, tables filled the streets outside the courtyard which were covered in pure white tablecloths that draped to the floor. The tables were covered in fruit, sandwiches, and cakes of all description, pies, sausage rolls, fish, bread and meat. Tea, coffee, decaffeinated of course and water, pure water which sparkled like a fountain as the light from the sun made it shine and sparkle, but did not warm it up. When the people opened the bottles, they were icy cold, just like they should be. There were squashes for the children and lemonade for all to drink.

There were even red and white bottles of non-alcoholic wine, plain bread, fillings of every nature people loved to eat. No one dared step forward, as of yet they had not been invited, not even the children moved despite their bulging eyes, dripping mouth's, sampling the delights before they reached their lips.

"We will take a short break," Mary said at last.

For the time she followed her Father, Mike was unaware of the intrusion into his body, mind and soul. Time stood still for everyone and the last question was wiped from tapes and people's minds. To the people of Earth, the question was never asked; such was the power of God.

"This is the time for all you to eat normal food; despite it being a grand day, we cannot allow you to fast; now I am here on Earth, I will eat with you."

She stood and walked to the nearest table which had been arranged for them. She thought to everyone around the world. "Please, let us pray and give thanks to God for the bounty before us. She continued with a prayer and then said grace and blessed the food and wine. Lifting Her arms wide apart, she invited everyone to eat and drink with her as a celebration of her coming to Earth.

CHAPTER 15

Mary mingled with her guests, shaking hands, listening to heart breaking stories and talking to children. An hour later, after everyone had feasted and rested, the tables disappeared as mysteriously as they appeared and this was happening all around the world even in countries where there was very little food, except here, the tables remained full of food and was as fresh as the time it was placed there. With the sun still at high noon, the entire Earth was in full sunlight and no matter where people looked, the sun was high in the sky looking down on the people below.

"After talking with the people gathered here, is there anything you would like to say to soothe any troubled thoughts?" Mike asked when he had taken his seat again opposite Mary.

"Yes Mike, it has been a wonderful time for me today and for those who are troubled, not just here but around the world, I can assure you they have all felt some form of relief. I am not here to sacrifice myself for your sins, that day is long gone. I am here to explain to the people where the bible was interpreted incorrectly and put the bible misgivings to right.

"I am not here to judge, or perform miracles on a daily basis or heal everyone who falls down and cuts their legs. However, those who go against me may come to wish they hadn't."

"Would that be a threat?" Mike asked quickly, not letting the quote end there.

"No, it is simply meant to tell those who are against God will find living with Lucifer is not as pleasing as living their final existence in Heaven," Mary replied with a warm smile.

Mike instantly recalled sayings when people had been bad, very bad and done very bad things against God, they would not go to Heaven.

"Can you elaborate on that? We have always assumed you go to one place or the other, up or down so to speak."

Mary grinned, reading Mike's thoughts.

"But," Mike continued. "The way you just said it, I assume providing no one goes against you, we will all be in Heaven after we die; am I correct?"

Mary paused; looked at her disciples and then to Mike. "There is much work to be done; it is up to the individual to make her or his own decisions whether to go up or down. It all depends on the way they act to others, live their lives and generally help in the community or in life.

"I know there are a lot of people out there, frightened to come out because they are either gay or a transvestite of one sex or the other. This subject has been dealt with in the past few days; but let me put it to you firmly in the way I wish it to be known now."

"That would clarify the whole mess, the entire world criticises these people, puts them down and shuns them from society when they can. Only a few really come out and get what they want from life. This generation, and I am sure future generations will never change their minds unless God intervenes in some way," Mike explained bringing Mary up to date with the feelings of many people on Earth.

He knew it would not make the slightest difference to their thoughts; after all, they did in the first place make people on Earth besides other planets in the galaxy or galaxies.

"The problem is, you discovered the truth far too late and it was never accepted by doctors and politicians alike. Because they were not protected by law like everyone else they where outcasts. People to be stoned, shunned and generally thought of as queers," she paused for a long time, letting Her statement take effect on the whole world.

Now everyone knew what they were calling these people and what they were doing to their lives.

"It was a slip up in the chromosome, a slip which could occur in any race. The trouble was, because you were treated as queer, they introverted into themselves, frightened even of their families or what people or their friends would say about them. It has only been during the last twenty years they have come out and told the world they actually exist."

"Will you make them normal again?" Mike asked.

"Who is to say what is normal; I will show you what life on the planet Triped, a planet which you will visit in the next five years, looks like."

There was a sudden gasp from the crowd that they were about to witness a real living extra terrestrial.

"The species on the planet Triped look like this."

An image of a three legged upright being, about 1.5 metres tall, with two heads and four arms stood looking out over a red sky, with two suns high above it. The being was pink in colour and didn't appear to wear any clothes.

"Is this person normal?" Mary asked.

"Compared to us, I suppose we might call it a freak," he replied.

"You are not normal to it; to Hariuam, that is his name, as close as a translation as can be, and to him, you are the freaks. On their planet, everyone looks like this, they also go around naked, the temperature allows them to live naturally. With you fully clothed, baking on their hot world it is definitely you who are not normal; it is you who are the freaks. How does it feel to be called a freak, knowing deep down you are made the way you are? Not only that, there is absolutely nothing you can do about it to change yourself."

"Bad, very bad, but surely they will listen to us; after all, we will have come from another planet?"

"Not really, you see, on their planet, due to a similar freak in their body evolution, there are people just as yourselves; they are shunned and called freaks of nature."

"Then it is different, I like everyone else would feel unjustified in being treated this way. It is very upsetting, you feel annoyed, as if you wish to strike out and shout I'm not a freak."

"And so do these people on Earth, they are not freaks, but human beings, trapped in a faulty body; it was not their fault; there is nothing they can do to alter it. It is time these people are taken for what they are; for once to be accepted in life without being stared at, or talked about behind their backs. They will be the stepping stones; the first steps in accepting different forms of life on other planets. Those people who cannot accept their own kind will never go forth and meet other alien life forms which live in the universe. They will be condemned to remain on Earth until such times as they repent most sincerely, the way they believe," she said firmly.

Mary gazed up to the stars that were now shinning through the dark blue sky; the sun had momentarily dimmed its brightness to allow the night sky to take form. Different stars and constellations could clearly be seen, as if magnified for the people of Earth. There were no clouds in the sky; giving the entire world's population a wonderful view of the night sky in the middle afternoon sun. Planets of all sizes and descriptions filled the sky, like a revolving picture of what space was really like, if the people of Earth would accept it.

"From this day forward I will allow these special people to be what their hearts desire, not what other people wish for them. When they have made their decision, their families will accept them for what they truly wish to be. They will have nothing to fear as they will be the real people they wish to be. They will do no harm to anyone, but if you fear them, you fear God.

"As you can see, the stars and planets out there in space, all these thousands of planets have life on them, species similar to yourselves and species so unfamiliar, you would never think a species possible. Many do not breathe oxygen but methane gas and even carbon dioxide and some breath through drinking liquid hydrogen. Just like these people, your people here must be accepted, or how else are you as a species to move forward to the stars to meet other life forms without feeling embarrassed and horror at the sight of them?"

"What exactly do you mean by, they will have their hearts' wish?" Mike asked.

"The transvestites and transsexuals in question will have the determination to come forward and change their sex. This will not be through an operation; they just have to decide what they wish to be, man or woman. When they have made their decision, they will be either male or female and their body will be the same. They will be accepted by their families, but their friends will have to accept them for what they wish to be and it is these people who will be put at risk for not travelling to the stars. Surely you should never mock what you haven't tried?"

"That is true," Mike agreed.

He recalled the moment he was transformed into a woman and the feelings he felt, secretly deep down inside. It was pleasant, he did like it, the clothes were different, not like the clothes of men, there were so many different colours to wear, so many different gowns and materials to make one feel elegant. No matter how expensive the suits were, they never made him feel like he had for those few short hours he was wearing female clothes.

"Governments throughout the world will acknowledge them as normal people; they will from this day forward, always be treated as such. This is my second law which will be given to you all.

"Treat others as you would wish to be treated, with respect and dignity, from this day forward, there will be a third gender, The Bisexual. These people will definitely travel to the stars and some will be on the first ship to show other species we accept diversity in species."

An enormous clap of thunder sounded overhead and the words Mary had just said became engraved on the floor behind them by the entrance to the church. The first law was put into every bible which had been made by God.

At the same instant, people's lives throughout the world that were affected this way, became less stressful and more tolerable. The people close to them understood them for what they wished to be as they walked into the open, many for the first time as God put their body to match their mind.

As politicians watched the screens, many were amazed as they too were changed to the sex they wished to be. Somehow, Mike knew what was happening throughout the world as scenes changed, so did people's lives. It was just as if he were watching a television show, Mike was amazed to see the number of people throughout the world affected by this secret they had kept all their lives.

When Mike returned to his senses, he looked dazed for a moment; he listened to John shouting in his ear. Something had happened to him although he was not aware of it, then he looked down, to see he was wearing a dress.

"I'm not a . . ."

"I know, but I made a point of you, to show the world here and now how people can change."

Mike looked at himself again and was a man once again and sighed deeply. He then recalled his earlier thoughts and wondered if the decision was made on that alone, so he pushed the thoughts to the back of his mind and listened to the voice in his ear telling him what to ask next.

"I'm sorry, I was taken aback for a moment, are there any further laws you will be giving us?" he asked bluntly.

"The old laws stand as they did thousands of years ago, the church will be free to marry divorced people, gay couples of either sex. The church will recognise gay people as well as the courts of this planet; people who have had a sex change will also have all their documents altered right now to the sex they have chosen.

"Governments throughout the world will no longer make war upon other governments, there will be no more hostile takeovers from governments and the government will be chosen by the people, not the people who wish to be in charge of a country or regime. From this day forward, all weapons of war will be disposed off, with the exception of laser systems which will demolish old and derelict buildings faster to build new homes for the people of this world.

"The Sabbath for all shall be a Sunday and will be a time of prayer. For those in work, the shop and factory managers will allow ten minutes of private prayer in their time. It is not much to ask, there is no need to fast or eat different foods on certain days of the week, times have changed, we will move with the times and it will continue to be so in the future."

There was another loud clap of thunder and a brilliant blue flash of lightning which engraved the words of the new laws onto the floor outside the church. Again the law was inscribed into the new bibles throughout the world.

"Will you be travelling to other parts of our world?" Mike asked.

"I will see every country and meet with many people; it is now a big world, much larger than when my brother came to visit your descendents. I will make many new disciples and they will all be given powers to help spread the word of God. They will explain what God stands for and what She wishes to be done. They will spread the word that God is feminine and loves everyone and this time, things will change and many miracles will take place."

"Talking of being feminine, this will obviously upset a lot of men, what will happen to those men who hit or mistreat their wife, women or children in general? I refer to rape which so often has gone unreported because of the threats by man against the woman, even the other way around."

"Any person, man or woman, hitting each other in temper, rape or humiliating them before others in an unjust way, I don't include what you call jokes by people who do this for a living, they will see what it is like to have it done to them. The change will not be just for a moment, but always.

"They will feel the anguish, hate and misery and be made to go through what women suffer; they will even bear children and be tied to at least six children by one man or another. Although it is unjust for women to spoken ill off, among men alone, they will be talked about. Everyone will know them by sight, for the colour of their hair will be. . . Bright green."

"Green?" Mike questioned.

"For the evil thoughts and jealously they have against women, which have cultivated their mind, they will never be allowed to dye their hair another colour, it can only be highlighted and never shaved off so they are bald and neither will a wig stay on their head."

"So men had better be careful in future how they treat women."

"Women have been ill treated far too long, it is their time to come forward, be strong for each other and live as they wish to live."

"When will you be starting your travels?"

Mary looked up to the dark sky and picked out a planet; as She thought about it, it became larger as too to its entire planetary system.

"You see that planet? It is called Epsilon Three, twelve light years from here, it will be the fourth planet the first ship will visit, the people there are very similar to you and their belief in God is failing. It will be up to the disciples to turn them back to God; this is what the people who go into space will be doing most of the time, passing on the word of God, besides seeing all the wondrous sights of the galaxy. Here," Mary raised her arm and brought another heavenly body into view.

"This is our nearest Black Hole, known to you as V404 Cygni and is 7,800 light years from Earth. The fifth ship to leave Earth will be very large indeed and be the first ship to travel to another galaxy, in fact this Black Hole will take the ship and its people to the Andromeda Galaxy where they will meet more people like yourselves and some people who don't. Now, to bring back the light." She waved her arm and the night sky left and the sun reappeared in its high noon position.

"To return to your question, I will be starting my tour of your planet the day after tomorrow, after I have spoken in great depth to my disciples and made many more, there are many people to see and places to visit."

"How will you be travelling around our planet?"

"In a coach, which is already at the Manson of Disciples, the building I mentioned earlier."

"You have just given it a name," Mike said noticing the addition.

"Yes, I have, the building is not quite yet completed and I am informed that is what the place will be called."

"Will your disciples meet there often?"

"It will be a place of refuge for all my disciples and me, a place of holy worship. It will be used for many things involving the running and organisation and trips while I am here," She said smiling warmly, her tones of importance were now gone.

Mike wondered if the conversation was about to come to an end, he glanced at his watch seeing it was four thirty. They had been going for a long time and yet, he was not tired, despite being up for almost a day.

"How many disciples will you have," Mike asked quickly, trying to cram in as many questions as possible.

"As many as it takes, the designers of the space ships which will take us to the stars are also my disciples, it will be their sole task to come up with bigger and better designs and make these ships available to your people."

"It sounds like the Ark in Noah's time," Mike replied.

"Yes, you could say that, but these are not ships of war or made of crude wood, they will be made from alloys and be able to pass the speed of light. These ships will travel not on water, but the seas of the universe. These are ships of peace; to spread the word of God through the Cosmos, so everyone will know what is happening and who God really is."

"How many ships will be built?" Mike asked.

"At least five hundred, in the first few years, then we will build bigger and faster ships to take more people to the far reaches of space. I will send large city ships through space, holding a million people to see the wonders of the universe and to spread the human race far and wide around the Cosmos."

"Will there be any ships for man to research the cosmos?"

"I will allow these to be constructed; they will be smaller and carry scientists of both sexes, for space was put there for all of you. Men will go with them, as there are many men who would like to see God's universe as well. Man however, is not destined to dominate the universe as he has dominated the Earth." Mary smiled a knowing smile that was neither evil nor discourteous; the reply had obviously come from high above.

"We have seen and heard that even today; Lucifer has tried his hardest to interrupt your special day, using men as his weapon. Does God or yourself have no faith in man at all?" he asked seriously, a question for men everywhere.

Mary looked him deep in the eyes, this was such an unbecoming question from Mike all of a sudden, what she saw she did not like.

His eyes glowed black, jet black. Lucifer had managed to infiltrate this man yet again sat opposite God's own daughter. Now Lucifer would pay dearly for what He had done. Then something happened which she was not expecting, Mike's eyes looked at her, gazed lovingly over her body and Mike's right hand gently reached out and took Mary's in his. A thought entered her head which did not come from Mike.

"I just wanted to see and touch you today, it has been so long, and you have grown into a beautiful woman my daughter."

"Will you please leave Mike's body, I cannot talk to you like this Father," She replied into His mind.

"Yes, continue your day, I will no longer interfere for the rest of today. Just remember what was said before all this started, for time is different here and I have been away for so long."

To ensure her Father was gone for good, Mary stood over Mike and placed her hands on the top of his head pressing hard, saying the Lord's Prayer over and over. Her disciples started to say the Lord's Prayer as well and as the chanting got louder, so more and more people around the world joined in; as if asking God for salvation for everyone and most of all, Mike, who had been taken over by Lucifer.

After a few minutes of praying, lightning bolts travelled around the Earth, striking down on people and hitting them to the ground, burning evil from their body and thoughts. When the lightning stopped, Mike dropped to knees and kissed the white marble floor. Slowly standing, he looked into Mary's eyes.

"I thank you from the bottom of my heart for my salvation, may you reign for many long years on our planet and travel the Cosmos in peace and harmony."

"Thank you Mike, may God always be with you, for you are now pure in heart like the many people I have freed around the globe. There is no longer an evil thought in your body and Lucifer will not trouble you ever again. As for what you secretly wished for some hours ago, you and your family will represent the press with others on the first ship to leave Earth and head for distant planets.

"Thank you Mary, I am very pleased and I assure you, I will do my job to the very best of my abilities. I thank you, from the bottom of my heart once again for this privilege," Mike replied.

"For those of you watching, what you have just witnessed is the demons which roam the Earth in search of salvation who get into people just to try and go home with the persons own soul have now been wiped from the Earth and they are all now peacefully back in Heaven where they wished to be. The prayers that were said have helped those demons on their journey and the last few are now entering St Peter's Gates and being welcomed back into the bosom of their families.

"All are now at peace in Heaven, no harm has come to anyone on Earth and the people affected are perfectly well, but now blessed with the inner spirit of God."

Both sat back down and Mary smiled at Mike, giving him a moment to catch his breath and realise what had happened to him. The pause gave John time to give Mike another question.

"God does have faith in man, but man was not meant to lead, to control women like he has done in the past. When man can accept woman as his equal and no longer treat her as his slave, then, and only then will he be allowed into space by himself. However, on saying this, there are many men around the globe who do treat women as their equal in the workplace and in the home and do their equal share of the chores and bringing up their children. Many of these men are frowned upon by their workmates and colleagues alike, but they still continue to help their wives, mothers and daughters alike. These men alone, who have shown us they care for and love their female companion as their equal, will be allowed into space, and there are enough men in the globe to be able to fill a number of ships and these men with their families, will be taken aboard the ships and sent with God's blessing into space."

"Can you please explain further what you mean by that, for those of us who have just been touched by your hands and filled with the Spirit of God already know what you mean, but others will not? For those men who do not fully understand what has just happened will need further clarification so everyone fully understands the importance of being equal," Mike asked thoughtfully.

Mary paused for a moment, as if pondering Mike's question. It was rational, but she knew She would have to answer it in a way the simplest of men could comprehend.

"Man has assumed he and God are one alike; man was made in the image of God, God took a bone from Adam's ribs and made Eve. This has given rise for man to think that God has wanted man to be in charge of the whole planet, when in the garden of Eden, as was wrote in the old bible, it was woman who ate the forbidden fruit and ruined a perfect world. Man assumed it was woman who flunked the test and therefore man was supreme and the dominant race.

"If man was left at home all day to do only housework, cook meals and generally do as a woman does, then he too, if he did this from day one, would have very little muscle. He would not be able to carry out the heavy work without the training and help which was needed for this type of work. It was the position women were put in, to wear dresses and be turned into slaves for man. In millions of years gone by, men and women wore identical clothes, they all wore their hair the same length in many countries; in many hot countries here on Earth, it is far wiser for men to wear a dress than trousers, because of the heat.

"Only a few centuries ago, men always had long hair, wore it permed, wore makeup, they also did very little work and thus had different muscles; in fact, many of these men never wanted muscles because of the delicate clothes they wore; even Kings and members of the governments. Yet now, many of you would say this is feminine, even your ancestral Kings were like it, as too Kings throughout the globe. Women were not meant to be slaves, but man's equal, originally this should have been a planet of women only, so it was the job of a woman to rule, not like man has ruled now," Mary explained.

"So what you are saying is; if man wishes to enter space, he will have to change his ways on how he treats women?"

"Yes, except those men who I have chosen, you included," she said for everyone to hear. She looked around at the people gathered about her, seeing the look's on their faces, knowing they were pleased to have her with them.

"It is coming to the time I must end this interview and discuss other matters with my disciples. However, I could not allow you all to leave without first advancing my hospitality to you all and giving you a final meal."

Once again, tables were filled as before to fill the streets, roads and parks wherever people were gathered. Those in their homes also had smaller versions of the food set up before them.

Raising her hands and saying grace, the children and parents walked casually to the tables and tucked in. Some thought about the starving in Africa and while Mary made her final rounds, one woman managed to take hold of Mary's hand and bring this very question up.

"Mary my Lady, I am an aid worker for the middle-east countries; you have provided this wondrous food here, what about the starving people there? Could you please do something for them too?" she begged, crying for the starving abroad.

"Catherine Patel, you will continue to work but no longer here. You will work in the new Aid Centre called Concord Aid, which is in Concord, the new capital of Israel, and be one of its leaders, for I know your heart is in God's work. You mean everything so sincerely, please, look into my eyes."

As Catherine wiped away her tears and looked into Mary's eyes, she saw pictures form of little children in a place she knew well. A place she had visited often and helped to distribute food and medicines; she had also seen a great number of children and adults die there from malnutrition. Normally it was a desolate place, nothing grew in the land and it was barren with no water to help purify it and bring life to the plants. Little children with fat bellies lay helpless in their mother's arms, while men did nothing, for there was nothing they could do, the armies had stripped the land and taken all their tools and money. All they could do was watch their families die.

The scene changed and the land was now fertile; with crops growing, more crops than they could possible eat. Children ran around fully clothed and fed; the flies were gone, water was flowing from proper taps and rivers irrigated the fields and crops. Schools and hospitals equal to any in Britain and America were there, small numbers of machinery and tools for the land, but it still needed tractors, trailers and combine harvesters to harvest the food that would soon be ripe enough to pick, good food to fill the empty bellies of the children and parents alike. There was so much that would still need to be done to make the place right, for those people to fully fend for themselves; grain silos, roads and railways would be needed to transport the food around the country, and to other countries where it was so badly needed.

Tables were filled with food and the children and adults were tucking into it just like they were in the UK. The scene changed to show other countries where the same scenes were being shown, Mary was indeed feeding the world on her special day.

Catherine bowed her head low. "Praise be to God and yourself Mary," she muttered quietly through more tears. "Praise be!" she sobbed in gratitude.

"Rise Catherine, disciple of heart, lead the people in your field, take your family with you and go to Concord Aid this evening; my personal jet will take you there, from this day forward, people will always wish to help you. Go and prepare the way for the helpers which will soon follow you from all nations, get the building ready to receive more aid than you ever believed possible from around the world," Mary told her.

"I will your holiness, but alas I only speak English and a little African and my French is not fluent."

"All my disciples and their families are blessed with multi lingual skills; you like other disciples have the power of mind reading, teleportation and other mind skills. Eat first with my disciples, especially first Lady Rocc, who will help you a great deal in the future. You will then depart on your mission; you already know all the arrangements and where you will you stay and what your rooms look like." Mary smiled as she gently touched her head for a moment then watched as Catherine understood fully all which she had to do.

Her clothes were changed to a pure white skirt suit and inside; she knew her body was no longer suffering from the aches and pains that had been with her for the past ten years as she worked in the hot climates. She didn't realise but other people saw a halo of light above her head for the duration of the feast. When she left, the halo left her as well, but she didn't walk through the crowd to get home; she simply disappeared, something people would have to get used to in the future.

Mike saw Mary for the last time to thank her on behalf of all the producers, camera crews and sound men and the help she had personally given him.

"This day will remain in my heart forever; it is truly a day I will never forget. Thank you once again and I look forward to meeting you on the first ship into space.

"I will not be travelling on that ship myself Mike, but I will be there to bless it and see it lift off on its special journey," Mary replied.

She touched his head and gave him a special thought. He smiled, on seeing his daughter, playing in Heaven with her friends; he knew she was not alone for she was with God and Mike's grandparents. Tears rolled down his face for her and his grief. A love he would never forget, he had been privileged to see her in Heaven.

"Sarah," Mike called.

His daughter turned, smiled and waved at him through an opaque mist and saw her beloved father, a man she had never forgot all the time she was in Heaven. Mike whispered to her, "I love you," in his thoughts. She blew him a kiss back, a childish kiss like she always did when he left for work and looked up to her bedroom window.

The mists folded over and the picture of Sarah and she disappeared, he looked to Mary through tears of joy. There were no words which could say how much he loved Sarah, and thanked her for the opportunity of seeing his daughter one more time, at peace, in the place he had hoped she would truly go.

"Before I finally go, there is one more set of gifts I will give you, the gift of multi lingual language, so you can speak fluently to anyone in the world and other worlds, in their native tongue and of course, write it down. I also gift you the use of teleportation and telepathy, so you can speak with other disciples so gifted."

"What does this mean for me?" Mike asked, unsure of why he was being given these gifts.

"It means, apart from the church ministers, you are the first male disciple outside of the church. I will see to it that I use you well and I will call upon you from time to time to tell the world of what I am doing or what I am about to do."

"Thank you Mary, I will do everything you ask of me," he replied. He bowed his head and taking her hand in his, kissed the back of her hand, noticing for the first time a large red ruby on a gold band on the third finger of her left hand. He kissed the stone and looked up into her eyes.

"Thank you," he thought.

"It was a pleasure talking to you Mike, thank you for making my first interview as easy as possible."

Mary turned and walked away from him, joining her group of disciples. Turning around to face the crowd, she looked around at her people and with her disciples behind and close to her, they all disappeared from view. As soon as they were gone, the sun returned to its late afternoon position in the sky and the light clouds returned over the town of Radstock. There was no rain, and the late afternoon sun was still warm enough to keep the children playing while the adults and teenagers talked among themselves about Mary and their new world that was about to start.

CHAPTER 16

When the group of disciples materialised a moment later, they were inside their new home, Mansion of Disciples, situated as Mary has said, at the bottom of Silbury Hill, with the massive burial hill behind it acting as back stop, so people could not see into the rear of the building. It was large, very large with over ten thousand family bedrooms suits with three bedrooms each, a study, lounge, bathroom and separate toilets, each fully fitted to a mortal's needs. Each room had a large 80" colour TV, computer terminals and computers already installed with all the latest functions and some no one had thought of. Each computer was voice activated and there was no possible way any virus could get through its multiple protective layers.

There were many rooms in the building that were off limits to everyone except Mary. There were twenty large conference rooms; fifty small prayer meeting rooms; eight libraries with all types of books people would like to read, from stories of romance to crime and SF, maths, to physics, all languages, spaceship design, engine design, theory of space travel and very complicated books on light speed and how to use and create faster than light particles. Each library contained at least five thousand books on religion and God. Then there were six very large libraries with all the old ancient scrolls which had been taken from the depths of the churches in Rome, England, Israel and other countries around the world. All these scrolls were now like new and able to be read by scholars from around the world. To help with the initial collecting and laying out of the scrolls and ancient books, all this work was done by angels chosen by God. Not all these angels had wings, but they were from Heaven, and no matter how anyone tried, they would not tell anyone from Earth what Heaven was like or what happened there, so all the staff for this and other special libraries, wore a pure white suit and looked like any other person from Earth, each able to speak any language they needed to.

There was also a special library for astronomy and the inhabited Universe and Multiverse. Books on every species alive, telling the reader their positions in the stars and how far from Earth they were and how far along in their evolution they had got. Pictures of Mary, Jesus and Mark filled the rooms and halls. There were also many pictures of other planets, their species and some lost pictures and new pictures from ancient artists who had still painted when they returned to Heaven, were hanging from the large pure white walls. Not all rooms were white; some were brightly painted for the children or guests of disciples.

There were four schoolrooms for the children and youngsters who would soon be filling the Mansion. The ten thousand rooms here were for the disciples and their families, further out in the grounds was another large building, just as grand, and this held two thousand bedrooms, offices and large conference rooms for the space engineers and designers with their families. God wanted all these people under one roof, so they could combine their knowledge and talk to each other in the same room, creating new ideas and making them work.

A third building, again very large and having eight thousand bedrooms, would be occupied by all the theologists from around the world who wished to be instructed in the word of God, to understand the new laws and read the ancient scrolls. Anyone who thought they would like to take this step in life would be asked by God if they would like to join Her group and be taken with their family to the Theology Mansion.

In America, Russia, India, Scotland and Kazakhstan were situated the five new spaceyards to build the first five massive ships that would travel into deep space. Once again each spaceport would have its own mansions where people would stay with their families and work and learn. There would be plenty of time to relax as well as many new Malls filled with goods for the people to spend their well earned money.

In the main Mansion of Disciples, were four music rooms, each with its own instruments and anyone could go there and learn to play any instrument they wished. There were also four recording studios, as God wished the people to be able to sell their music to the people of Earth and bring in money to show everyone this was going to be a big business and help the world see how a big business could be run without threatening people and making employees feel ill treated.

There were thirty games rooms filled all forms of computer games, pool, snooker, chess and other games, and outside, tennis courts, football pitches and even a rugby pitch. In the summer they would change the games to cricket and other summer outdoor sports, including archery and golf. There were swimming pools and every form of entertainment anyone could wish for. A train ferried people to four local towns where new shopping Malls had been placed, filled to brim with merchandise for people to spend their money on.

Only one thing was asked of the people who worked and lived in the Mansions, each day they were to pray for ten minutes in the morning before they set off for the dining halls and had their breakfast. Before they ate, everyone, children included would say grace, at ten in the evening, the people would say their last prayers of the night; it was not much to ask of anyone in this new world, and everyone did as God asked, especially as now they knew God really did exist; even though She was female.

The kitchens were colossal, as too the staff who worked there, in the gardens were numerous pools for fish and birds had to be fed every day. The wild animals also had to be fed and cared for and there were no fences surrounding them and quite often a fox would come up to a person and allow them to smooth the animal and it would purr in excitement at being here in God's garden on Earth.

As like many of the jobs in the Mansion, they were filled by people from Heaven and would continue to do so until people from Earth took up the jobs now being advertised in every jobcentre throughout the world. This was to be a very big business and would be world-wide.

There were also fifty offices just for sorting the people who were coming to the Mansion by train, car or helicopter, each person would need a room and a place to work, once again, to ease this burden, all the jobs were undertaken by people from Heaven who seemed very pleased to be helping do something to promote their true God.

As the families of the disciples unpacked their belongings and examined their rooms and the layouts, they were each given a mobile phone which had a floor plan of the mansion so they would know where they were and where they had to go.

There was a great deal to do and the families knew the disciples had their own jobs to do and would not be able to help them move, but the people found themselves surrounded by helping angels and they made the people and children feel at ease and broke down the silence and curiosity of this very large house.

"What is behind those doors?" Bishop King asked inquisitively as the disciples were shown around the mansion by Mary.

"Rooms designated with a silver star belong to the angels and people from Heaven who have access to the mansion. The rooms to you would be open space, with nothing inside, it would be like looking down from the top of a skyscraper which was floating high in the stratosphere, attached to nothing, resembling nothing; should you inadvertently look in as an angel was leaving or entering the room.

"To my helpers, this is their access back to Heaven. If you or anyone else, tried to open the door, the handle and door will not turn, no matter what force is applied to it."

"What about you, are you allowed into these rooms?" Bishop Meadow asked.

"I wouldn't enter these private chambers unless invited, so the answer to that is no, however, I do have my own room which is accessible to Heaven, something my brother Jesus did not have.

"Do the people from Heaven put themselves in any danger working down here on Earth? I mean, old memories, the chance of bumping into one of their family members?" Archbishop Langley asked.

"The helpers have their families in Heaven; they passed on many hundreds or thousands of years ago but have kept up the technology as we progress in Heaven. Of course we progress much faster than you, it gives the soul something to do."

"When we see them, will they be whole, fully formed so to speak? I don't mean to sound rude, but I have never seen a spirit before except of course yourself. I assume you live here as a mortal, like your brother Jesus before you?" First Lady Rocc asked.

"Yes, we have learned from our mistakes, we know now the power Lucifer has over your people and we have not underestimated his strength; I am therefore not completely mortal, although I take on a mortal form. I cannot be killed in any way, I can eat, drink and a have form, a body which is not translucent. Our helpers here have taken on a mortal form so as not to alarm anyone; however, they cannot leave the boundary of this mansion and its land. They cannot be killed or injured; they will all wear plain clothes, white of course so you know who they are. Helpers in certain areas will wear a coloured badge so you know where they are working. The people who feed the wild animals will all wear clothes to allow them to fit in with their surroundings and not un-nerve the animals. The grounds people will also wear clothes befitting their jobs."

"What about Lucifer, how do you intend to stop Him penetrating your stronghold and any other stronghold around the world?" First Lady Rocc asked.

Although she knew this woman could not be affected by Lucifer or His agents, Mary double checked everyone in the group first by looking into their eyes, bringing out any lies or impurities.

Mother has seen to it, there is an elaborate security screen surrounding this and every other complex God has created which is all controlled from Heaven. So, we have no fears where this or any other complex is concerned. Although to all intense and purpose this huge mansion is at the bottom of Silbury Hill, it is as you see very large indeed, this would encompass Silbury Hill a hundred times over and therefore is in another dimension that can be seen from Earth. The only way in is by one road and a special corridor for our own helicopter pilots. Likewise the trains and buses that come and go also travel through this dimension to Earth. Lucifer and His agents cannot gain access to this dimension, so we are all safe."

"That is very clever," First Lady Rocc said smiling with intense pleasure that such a building could not be penetrated.

"But I thought the building could be seen on Earth as you approach it?" Bishop King asked, sounding slightly confused.

"Many things which are made in Heaven may be seen on Earth, if it is the will of God, then it will be so. Now we will go over our plans, then you will have time to make your preparations and a tour of the mansion later."

They followed Mary to a very large, spacious room with easy chairs arranged in a circle with a coffee table in the centre already laid with coffee and sandwiches.

"Please; take a seat and help yourselves to food and drink," Mary instructed.

The room was decorated in a pale peach colour, with doors leading to a wide veranda which overlooked the massive gardens below. Further out, past small shrubs were very large, lush green lawns with trees and more shrubs, flowers and paths which were beautifully kept. It looked as if the place had been built a hundred years ago with the size of the trees and the amount of plants in full bloom.

The peach colour carpet was very thick, pictures of places in Heaven were placed on the walls, photographs of great inventors, men and women who had shown themselves well above others on Earth in a time gone by were also on the walls. On the ceiling, embossed forms of angels blowing their trumpets while flying through the air were in each corner. Around the centre light piece was a very elegant gold shaped heart, covered in tiny lamps which lit the room evenly.

"Before we commence, there is one more question I would like to ask if you please," Bishop King said, blushing slightly. Mary looked at her in silence, as if to say, I'm listening.

"It is for the children who will come with us really, do you have television, music centres and computers here?"

She felt very embarrassed having not seen her own room yet and now she realised it was a very trivial question, but she was a mother like other women here. She suddenly felt as if she wanted the ground the open up and swallow her for even putting the question forward.

Mary laughed aloud for the first time that day; she had smiled a lot, but now she was actually laughing, a sign she was human in some way.

"Christine my dear, all of you, television, music centres, DVD players, computers, what a question. Of course you are used to them on Earth and you use them in your daily tasks and of course, your children will need to use them in schools. You watched television to while away the weary hours and catch up on news throughout the world. In Heaven we have many hours to catch up on, it is not all work and no play, here you have the same recreations as on Earth, and a few more thrown in for good measure. You have a choice of all the television stations in the world, music of any type from our extensive libraries in CD's, DVD's, or laser. Everything is stored on computers and you can upload anything you like from the computers to your own computers in your rooms and watch what you like on TV or listen to music on MP7 players. These are more advanced players than you are used to you, it's the latest design in Heaven. It will allow you to listen to music, watch television in 2D or 3D, talk to your friends and see the inhabited planets in our galaxy.

"There are hundreds of cartoon channels which are in Heaven, although some of these are restricted. There will be continual updates in the design and building of the spaceships and when they are being constructed, every moment will be recorded by our camera teams who of course will not be seen. So to answer your question, yes, there are televisions in every room and your children will be able to watch whatever they like, and I hope, learn a great deal."

"Then people who see spirits are really seeing people from Heaven?" Bishop Meadow questioned.

"Yes, in your rooms, which you will be shown to after this meeting, if you turn to channel 100, you may have all your individual questions answered on any subject, except of course, Heaven itself. You will have to wait until you get there to discover what it's truly like, however, I can promise you one thing, you will not be disappointed and when you do traverse the leap to Heaven, you will feel no pain at all. This however, will not be happening to you for many; many years to come, at least one to two thousand years as you have a great deal of work to accomplish. There will then be ample time for you to catch up with old friends I promise you," Mary replied with a smile on her face. "There is one final point I shall make to put all your minds at rest about your families. They are all here with their belongings and settling into their rooms as we speak. Everyone is happy and there was nothing left behind, they know where you are and now you know their whereabouts."

"That's a relief, I was beginning to wonder what was happening with them," Bishop Meadow replied.

"Your new posts," Mary said, pouring coffee for them all and indicating for them to help themselves to the food on the table.

"Your new posts will be in the first instance, to learn from me, so you will follow me wherever I go for the first month. In that time, you will understand enough to be able to continue to spread the word of God wherever you go here on Earth or out in deepest space."

"Will you be travelling into space?" First Lady Rocc asked.

"No Jean," Mary replied using her first name. "Earth will be my domain, as it is Mother's project so to speak. She wishes me to visit every country in the world and sort out the troubles on the planet while there still is a planet to love and care for."

"We were heading for another world war I know, but I didn't realise we were that close," First Lady Rocc admitted.

"At times your leaders and some of your generals were bent on blowing this planet up with a situation of no winner at all. The world leaders would have killed their own people just to save their faces, what they saw as justice and peace, did not necessary mean the same to the other world leaders. None of them could meet each others' terms; they could not live in harmony as we do in Heaven.

"The plans have changed slightly; all of you, with your families of course will go forth into space and spread the word of God, with three lead disciples per ship. You will take with you over two hundred other disciples, and some of them will remain on the planets you visit and you will also make new disciples on the planets, which I will explain in detail when that time comes. The captain of the ship will be in control but you will have the overall word. You will say where he has to go and the duration of the time you spend on any given planet is up to you entirely. You will not, under any circumstances, leave the planet early or only half completing your mission as we want the word to really get through this time."

"If you will not be with us in space, how are we to know which planet to visit first and in the following order?"Archbishop Langley asked.

Now was his time, he would either be told outright he would not be going into space, or be talked to afterwards and given the bad news. Deep down he hoped he would travel to the stars; it would be a challenge to meet all challenges.

"Ah Robert, I was wondering when you would ask this question," she said using his first name, like she had started to do with the disciples. "God will tell you where to go in the form of a letter and your ship's captain will have a copy. He will then not assume you are cheating or keeping them out in space for any longer than is necessary; it will just materialise as you did here when the time is right you will be told by God when to expect the letter.

"The other people on the ship with you will carry out experiments on the planet and start to form relations with the governments while the captain and some scientists leave you on the planet and map the star systems and learn more about the area. If he finds objects of great interest you might wish to see, he will be able to take you to them and even people from the planet if you think it will help you get the word across. These will be the only times you will be allowed to leave the planet for short periods to show people things God has made."

"Can you tell me if I will be on one of the first ships?" Archbishop Langley asked full of trepidation, as his question had not been fully answered for him to know if he would be allowed into space or not.

"Yes Robert, you will most definitely be on the second ship; you twenty disciples will all be on the first five ships which will leave within a month of each other. By then you will each have at least two hundred new disciples under you."

"Thank you," he replied sighing deeply. "Thank you very much indeed for making this clear for all of us."

Everyone could see how happy and releaved he was; knowing at last he was going into space and so soon. Mary smiled at him and everyone else in turn, seeing the joy in their faces at being given such a huge and dangerous job. It was so exciting and thrilling, to go as James Tiberius Kirk, Jean-Luke Piccard from the Star Trek series, *'Where no man has gone before,'* into the unknown. Space itself; on a quest from God, to go in Her name. Mary paused for a moment longer in silence before continuing, giving them news they were not expecting.

"The time you have to get your own planet in order before you leave is only fifteen months. That is when the first five ships will be ready to leave. The plans for these ships, including the engines will be completed by tomorrow. They will all be built at these locations," Mary said pointing to the map that appeared on the wall behind them.

The shipyards have already been constructed which will be a tremendous assistance to the building engineers. Not only that, they will have antigravity devices to help lift the huge sections of metal onto the ships and move the large engines and generators into position. This will cut down the amount of time in their construction.

"Not only will you be working on Earth to get your people ready here, you will also have to get used to working in space and transporting huge amounts of material into orbit. They will also find for the first five ships, that the hulls have been laid; the engine components have been made and laid out for the engineers to observe and make more off."

"It seems this was planned some time ago," Archbishop Langley stated, taking a sip of the delightful coffee.

"This has been in the planning for many years, we have throughout the globe placed many people with gifts in their minds, but man has been slow to recognise these facilities and act upon them. Instead, these people are tested and put down as often as they can be. They are called freaks and thus for this reason, many others who were gifted, often remain silent in the light of what has happened to their colleagues throughout the world."

"How large are these ships going to be?" Archbishop Langley asked with bright eyes. He had followed many of the early Apollo missions and the thought of astronaut training had not entered his mind before. Their ships were small and cramped, spending months or even years aboard one of those ships with a small crew was not what he had envisaged for the latter days of his life, not that they would be the latter days any more, it was more like the mid years; even though it would be exciting to go.

Mary smiled upon reading Robert's thoughts; she had read the minds of everyone before her. They all had the same thoughts in their head's, they all had the same picture of a small craft, cramped and floating in space during their long voyage. The gigantic ships in films were purely science fiction, they would never fly due to their pure size and weight alone. They were also being built on the ground and everyone knew that huge rocket boosters had to be used to get even a small payload of seven tonnes into orbit. They knew if the ships were being constructed on Earth then they had to be small to get them into orbit, also, they were building five in fifteen months so they could not be that big. An enormous ship would take years to make, then it would have to get into space. They thought the figures Mike had given them earlier must have been wrong in the excitement of the day, or at least they were exaggerated.

"The ship," Mary said breaking the long silence, "will each carry a crew of," she paused again, looking at the eager faces facing her. "Twenty thousand personnel with their families and will be called ARKS; many of these people will be scientists, there will be six hundred disciples on each ship of which you will be in charge of their groups, how you split them up will be up to you. There will also be a hundred staff which will represent the world Government and will be in charge of making relationships with the people on the other planets and Earth. They will make trade agreements between our planets and as we meet other species, they too will be included in the trade agreement. Then there is the crew to fly and maintain the ship; they will also be in charge of taking people to the surface, helping others and generally doing what they are needed to do. Another team of reporters, headed by our friend Mike Adams will collect pictures and make reports on what you are doing on each planet. He will also be in charge of sending back pictures of what they see in deep space back to Earth for everyone to see. The ship will also carry a team of doctors, surgeons and nurses, in case of accidents on the way and each doctor will be given the gift of telepathy and multilanguage speaking. They will also help cure people on the planets they visit . . ."

"I apologise for interrupting, but did I hear you correctly, or is my mind deceiving me?" Bishop King asked.

"You heard me correctly; there will be over twenty thousand people on the ship and you will not be floating weightless aboard it, it will have full Earth gravity."

"Then the ships will be big, but how big?" Bishop King asked.

"The first four ships which you will be on will be three kilometres long, one and a half kilometres wide and one and half kilometres high, with fast lifts and pathways to different sections around the ship. Everyone will have a small powered bike to get around, but many of you will only need to remain in some places. There will be no need for astronaut training and nobody will need to wear a spacesuit or go walking about in space. The people who need to do this will be the crew and they will have new spacesuits that are nowhere near as cumbersome as they are now. They too will need no training as they will be given the gifts of knowing how the ship operates and it will have everything that is needed to repair it while on its journey through space.

"Let me explain, science fiction is about is about to become science fact, reality. The ships built on Earth will lift off from the surface despite their gigantic size. They will carry their own gardens, food factories, farms and supplies for all the people and their families. The ships will also carry some animals that will live in farms and the farmers will care for them and the food that is grown in special fields. We will not be sending just one person from the family; it will be the chance for everyone in the family to go out to the stars and be the first humans to interact with new species and see the wonders of God.

"Should relationships bloom while people are on planets, they will be allowed to remain there and be given a shuttle and communications equipment so they can contact one of the larger ships and return to us if they so wish. The interplanetary shuttle will be quite large and the species will be given two space yachts to get them to the next stellar system if they have no spaceships of their own. Many of the trade delegations will need to remain on the planets they visit to start to trade with Earth, so there will be many more people and disciples to continue your work on the planets. You will also be allowed to take species from the planets with you and also, have their trade delegates flown back to Earth."

"How will we get the people back to Earth if our ship is to fly deeper into space?" First Lady Rocc asked.

"I have a few designers working on a special ship which will be much smaller, let's call it a space yacht, a quarter of a kilometre long, half that wide and high. It travels extremely fast and is well equipped with the entire latest feature for luxury travel. When you are ready to send people back to Earth, we will send the ship to you and it will return the delegates to us. During the next three years, I will have another building constructed especially for trade delegation. My designers are also working on this project too and I have yet to decide where to put this complex, but it will be built.

"As time moves on, we will set up outposts where people will live in space on what you call space stations, although they will be large and self supporting. Your people are about to start on a wonderful journey, and many millions of people will be needed for this project from all over the world. They will be able to contact planets and sort out any problems and get help where it is needed. As your people get better at building the large ships, some smaller scientific ships will be constructed and these will take researchers into space. Normal space travel exploration would have ended up this way, but it would have taken you another 800 years, I have just speeded up this procedure with help of course and permission from my Mother."

"I never thought this would be possible in my lifetime," Bishop Huntley sighed, holding his hands together in front of him.

"There are many things woman and man feel it is impossible to do, but for those who are willing to try hard enough, anything is possible when they put their heart and soul into it," Mary replied and looked reassuringly at her disciples.

"For the remainder of this week and the next two weeks, you will accompany me; we will concentrate on visiting churches of all denominations, staying a short time talking to ministers and people in the church. For this initial period, when it seems the best way into people's hearts is through miracles and cures, that is what we will do.

"You have already sent some of your disciples to hospitals and prisons throughout this country, they will do God's work and heal and free the people, ensuring they will never again repeat their crimes. During the following two weeks, you will each take a few churches and start to spread the word of God yourselves and you too will be able to perform small miracles, heal people, and give them spiritual healing. You will all during this time pick out more disciples and people to take with you into space. Don't worry, you will see the good in them and what position they will they take. Bishop King, before you depart on your mission, as all of you will, I will be forming a new tower of authority for the United Church."

"The United Church?" Archbishop Langley asked.

"This will be the new church that will combine all faiths together and be headed by the one True God."

"Are we allowed to know how long you will be staying with us?" First Lady Rocc asked.

"I will be here for as long as I am needed; however, you will all out-stay me by several hundred years," she replied smiling affectionately.

"The church will have its own authority and soon, many churches included in the so called Church of England's Christianity, will be cancelled and brought into the new church. That is why I will have to talk with leaders of all ministries."

Silence filled the room as they contemplated what was wrong with their religion as it was, God had picked their church, their leaders were positive they had taught their people correctly, that theirs was correct and all others wrong, in a minor way or completely in all its aspects of belief.

Mary looked around at the wondering faces before her; she melted with their thoughts without their knowledge, to see what they were thinking. Her own thoughts were not that far off.

"I see you are all disappointed my children, you seem to forget, your church is based upon a God who is male. A male God has dominated your church for two thousand years, therefore you will laterally think in male ways. Everything you do in your church will be male orientated, nothing female. You do not have the feminine touch in the way you wish the order of the service to run. To be blunt, you may as well be praying to a Golden Lamb for all the good it will do my congregation."

The disciples were instantly shocked at her remarks, but there was worse to come. They sensed it and prepared themselves as best they could.

"Yours is by far the worst form of belief among all the faiths, you have allowed a male to dominate your lives both in the church and at home. To all intense and purposes, you in your unknown state were worshiping Lucifer since the death of Christ. He never wanted this, but Lucifer saw His chance when Jesus was nailed to the cross. That part is reported well, but his final words were muffled. He never cried out to his Father, but his Mother. His words to his Father were, "Do not punish My people for what they do not understand!"

"You see, Jesus was begging Lucifer not to intervene and destroy his Mother's creation and dream. Lucifer quickly spoke to the writers of that day and had the record changed to how He wished it to be. Making it appear as though a man was God Almighty. From that day forward; with the disciples soon after killed and mocked, they did not do as Jesus commanded of them. From that moment on, Lucifer was in charge with anyone knowing what had really happened killed. He was then in a place He was forbidden to go, God's Holy Church."

"Then we truly are sorry, you are right, a new church does need to be forged. Will we still have archbishops and the like?" Bishop Huntley asked.

"No, it will be First, Second and Third Lady, Mothers and Fathers of the faith and so forth. The complete list of titles, together with all your new titles are in your rooms. There is also a new format for the service and the way the church is to be decorated.

"When the New Church commences in two weeks time, all the old churches will be demolished and rebuilt to God's design. She will do this Herself and it will be in the darkness of night and the churches will not be able to be seen being demolished or rebuilt by anyone until the following morning. This will happen right around the world, to every form of church," Mary explained.

"It all sounds intriguing and very mystical will we get a chance to see a prototype of the new church before this happens?" First Lady Rocc asked.

"Only my Mother can answer that question, I cannot speak for Her. Now, I suggest you all go to your rooms and study the documents you have to read. All the arrangements have been made for the following days and you will do well to remember it by heart, like everything else you will have to do and I can assure you, you will have no trouble recalling what you remember, this is why I have given you your powers. We will all dine at eight this evening with your families as well. Now, your guide waits outside the door, a tour of the mansion will take place after dinner," she said and standing up smiled, then bowed her head. They all stood and bowed back and Mary disappeared from the room.

The door opened and a woman wearing a white suit and a red robe stepped into the room. She spoke warmly and softly; almost as if she were not in the room with them.

"If you will be so kind as to follow me please I will show you where the dining room is and then take you to your rooms which are on the eighth floor."

They left the room walking along the well lit passages over thick piled, beige carpet to the dining room then to a bank of six lifts which were very large and elegant, which took them up to the eighth floor where their rooms were.

Dropping each person off in adjacent suits, the guide put a picture of the mansion layout into their heads and indicated the route back to the dining room. She didn't speak after her first words and no one spoke on their departure.

First Lady Rocc entered her room and closed the door behind her. It was a very large room, much larger than she expected. It was carpet was beige, thick piled, with a three piece suite, coffee table, large wall 100 inch wall screen with the remotes for it on the coffee table. She heard music coming from another room and walked through the first door on her left.

"Mum," Jasmine shouted at the top of her voice and ran to her hugging her tight. Jasmine was fourteen and her daughter, Ian was sixteen, and entered through another door a moment later and hugged his mother. Leonard, her husband entered the room from the same door Jean had.

"Jean; I'm so pleased to see you, we didn't know what was happening, we were told to collect our clothes together and then we were taken to this place. We're all fine and have been told what is happening, I'm a little concerned about the kids schools, but we have been told it will all be sorted over the next few days. We saw you on television, are you okay?"

"Yes I'm fine," she replied and hugged and kissed her husband. "Why don't you show me around?"

She was shown around the rooms, three bedrooms all very large and each room had a large wall screen, desk, computer and music station. Each room had an ensuite bathroom, bed, two armchairs and two desk chairs for the desk. There was a very small kitchen where they could make hot drinks if needed, a fridge and microwave for when they needed to work late in their rooms. There was a study with a library filled with various books on theology and science; there were also Kindle books for each person and the very latest laptops and tablets; one for each member of the family. There was another study for Leonard as he would now be working for the mansion, but as of yet, he didn't know what he would be doing.

As Jean and Leonard sat down in the lounge she said, "I could really do with a nice hot cappuccino." A moment later there were two cappuccinos on the coffee table before them and they both looked at each other stunned and laughed.

"This is good service," Leonard said still laughing and picked up his mug. After they had their drinks, Jean looked at her husband, they kissed and embraced and both knew she had work to do.

"Do you mind doing the unpacking and seeing to kids; I have a lot of books and paperwork to go through before dinner at eight?" she asked.

"Don't worry, I have to do the same, loads of work to get through, all the unpacking is done and the kids are fine. They have their work to do as well and we will join you at dinner tonight."

"I'm sorry; I've taken you away from your friends and colleagues at work."

"Are you mad? We're going into space to pass on the word of God, we'll have the time of our lives, the kids are really looking forward to it, they will be among the first children to travel in space and go outside our solar system. They've both unpacked, done their first set of lessons and have been playing on the game systems they've given us. Don't worry; they're both as excited as I am."

"Really?"

"Yes, so get on with your work my love, don't worry about us, we'll be fine."

By the time 7:45 came, Jean came out of her reading and looked down at the pile of sixteen books she had read cover to cover and on her Kindle reader; she had read a thousand pages of notes for the trips ahead of her.

She quickly went to their bedroom, changed into a different colour dress and joined her family in the lounge, which now had family photos on the wall and a very large photo of Concord Aid.

Her children were clean and dressed for a proper dinner in their best clothes and looked very tidy, much tidier than she ever seen them before. They were also very polite and even bowed and curtsied to their mother. Leonard was wearing a dinner jacket and a white shirt which she knew he never owned or would wear before. Holding his arm out for her to take, the family opened the door and walked into the corridor, meeting the other families who were leaving their rooms and making their way to the lifts.

They entered the dining room a few minutes to eight, talked among themselves and everyone was wondering what Mary would wear. None of the women wore makeup through lack of time more than anything. Mary entered the dining room a few moments later wearing a pale green dress showing her knees; she wore tights, like most women, low healed green shoes to match her dress and wore her long hair up and was wearing some very light makeup. She looked just as beautiful as she did in her plain, long white robe. As the women looked at each other they admitted Mary looked just as beautiful as before.

"Please, everyone, sit, make yourselves at home and feel relaxed; this is the first time I have been able to welcome your families as well, I must say I am very pleased to meet you all and I hope you enjoy your stay and work here, as for the children, their studies will continue, only a little faster than you are accustomed to and you will have a few new subjects, astronomy, and space research being two of them. Of course, to help you with this you will all have access to the large twenty foot reflector telescope in the garden and each of you will have your own ten inch reflector telescopes with cameras to take, I hope some fascinating photos of the heavens which I will personally wish to see."

"Really?" Jane Bishop Meadow's daughter who was 17 asked full of excitement.

"I will be taking an interest in all of you, after all, you will be representing me in space; of course I want to know how you feel and what you see."

Everyone looked at the table of food that was spread out before them; it looked so tempting they wanted to dive in but waited, even the children, which was so out of character for them, for Mary to say grace which they all, even the children, joined in with. All the parents were amazed at the change in their children who had suddenly grown up and respected people and had manners to be proud of.

"The gift of keeping you to correct weight will also apply to everyone here, of course for the children, as they grow, so they will need to put on weight and this will be taken into consideration. Please, eat your fill, enjoy the evening, this will be our first night together and I want to make it very special for you."

With that said, a group of people appeared at the back of the room, two violinists, a harp player, double base and two people who played the cello commenced playing their instruments which sounded beautiful.

As they started to eat, a few people were talking across the table and as Bishop King was sitting next to Mary with her family, she decided to ask Mary a question.

"Mary, we assumed the clothes we were dressed in earlier today were a form of uniform, yet here we are, dressed casual and you are even wearing makeup."

Yes," she replied smiling. "It was thought as there were several members of the press and other magazines waiting to see me, the one thing they would want to write about was what I was wearing, which is why you and I were dressed very simply. From now on, and this applies to everyone here, you may wear what you wish until the New Church garments are ready to be unveiled. If you wish, while we are out and about, we could all wear the white suits or dresses. I will do as you wish, what do you think, you have been here longer than I?"

Everyone turned to face Mary on hearing her words and thoughts. Thinking quickly between each other as Bishop King brought the subject up, First Lady Rocc suggested it should be her to give Mary their answer.

"We all think, as we represent God and God is Holy, we should continue to wear the white suits or dresses, a couple have suggested we could wear a skirt and jacket and have a white hand bag to hold our bibles and perhaps a little makeup. We could also carry a smaller version of the tablet you have given us where we can make our own notes as we travel with you," Bishop King suggested.

"This is what I wanted from you, for you to question me, instruct me in your ways, as you know them best. Also, if you want something then you only have to ask, it is a good idea to have the smaller tablet, I'll have them sent to your rooms, again, for the husbands and wives of the disciples and your children, should you need or think we could alter something here, or there is something you would like to do, then ask any of the staff or indeed myself when you see me walking through the building which I will be doing a lot, as I want to get to see all of you and how you work here," Mary explained.

With the meal over, the children and spouses of the disciples moved to a large games room where there were large billiard, pool and snooker tables, computer games, chess boards with all different forms of pieces and some boards where the pieces moved themselves and fought on the board when a piece came to be taken. There were also three dimensional chess boards and hundreds of other board games. The large televisions were all connected to different game systems and there were thousands of games to choose from some not yet played or brought to Earth. While the groups of people got to know each other and talk over what had happened during the day and to their partners, the disciples were taken by Mary to another lounge where they talked over coffee and cappuccino, served by waitresses who had longed to help in the mansion. When everyone was settled, Bishop Meadow asked the first question.

"With our children and partners here, what will they do about their jobs and our children do about their schooling?"

"I am so pleased you have brought this up. Your partners have already been told what is to happen to them, and they have told your children what will happen but you have not had the time to catch up. Your partners will help work in the mansion for part of the day and then they will be taught what they will be doing aboard the ships when you travel into space. Some will learn astronavigation, others to fly shuttles, some will learn the art of trade and some will become doctors and surgeons. Everyone will have a job to do on the ship and this will help pass the time. Once on the giant ships, it will take a few months to get to the planets, you will not leave Earth and suddenly arrive at your destination, I want you to be able to see the wonders of the universe and some will help Mike and his team photograph the Cosmos.

"As for your children, they will all need educating and they will attend a school room here where they will study space, maths, multi language skills, speaking with aliens, astronomy, like I said earlier, and they too will have jobs aboard the ships. The children will also study advanced physics and spaceship engineering, botany and a number of other classes including Religion and farming: they will all be taught about the new church and how it will work. Some will become young disciples, these children, teenagers, will help you carry out your work and speak to the children when you arrive at your destinations. They will of course all be helped by being given a very good photographic memory and instant recall. Does this answer most of your questions regarding your families and what will happen to them?" Mary asked.

"What about the jobs they had and our homes?" First lady Rocc asked.

"Your homes have already been given to others who are right now moving into them. The jobs you all once had have also been taken by others and as for your larger families, you can speak to them anytime using the computers in your rooms and the computers that have been given to your family members, they will be able to see you on their large televisions and you can see them. If needed for any particular reason, your own parents could join you if both parties wished this to happen.

"As for the other disciples you made yesterday, when their current work is completed, they will join us here, but I have given them enough tasks to keep them busy for the next two weeks."

Everyone smiled, for although they were not close friends before their change of career, they had certainly become close friends now. Everyone was much happier and resumed the talks with greater enthusiasm.

As their talks on the next few days' routine ended, there was a loud knock at the door and a man entered the room dressed in a black suit, with a gleaming white shirt. He carried a silver tray and on it was an envelope which he handed to Mary and then left.

"That was Geeves, or so he likes to be called, he waits on everyone he can in Heaven. It was his pride and joy job on Earth and he continues to be a butler in his afterlife job," Mary explained.

"Does he have to do this for all eternity?" Archbishop Langley asked.

"Oh no!" Mary laughed. "You can do what you like for as long as you wish; you can even teach others to do a job like his. In fact, he has a small school of twenty five thousand pupils which we are all very proud of."

"Twenty five thousand," Bishop Huntley mimicked astonished.

"Not all at the same time of course," Mary added and smiled.

The group were astonished, then they thought of the number of deaths that had occurred on Earth, then there were the other planets throughout the universe, which had not even been discovered yet.

Mary read the note and looked at her disciples. "It seems our plans are to change slightly," she said gaining everyone's attention.

"King William begs an audience with us at Buckingham Palace at 10.00 in the morning. We will talk with the head of the church and some other leaders including the Pope, the Prime Minister and some of his cabinet. The Presidents of France and America will also be in attendance."

"That is some invitation, are we going?" First Lady Rocc asked seriously.

"Of course, my brother was once lured into a den of people just like this. This time however, we are forewarned and will be on our guard; times have changed and we are more prepared than before to succeed; even if it means wiping out the Head of State and the entire political structure; not just for this country, but the entire world."

Nobody asked what that would entail, but they all knew she was serious and God was behind her. They cast their minds' back to the stories of Jesus in Rome, they were not good times and Jesus soon met his fait accompli.

"The coach is outside and we will leave at 9.30 am sharp tomorrow morning. We will meet it in London and the coach will depart without us at 05.00 to draw away any attention."

"Will it leave by the normal route to the mansion?" Bishop Meadow asked.

"No, it is not a normal coach which you will find on Earth; it too has similar capabilities as yourselves. It will be seen in several different counties before it arrives in London. This means we have a problem I was not hoping to bring upon us yet. We will have a short thirty break for you to meet again with your families in the games room and then we will regroup here as you will need all your concentration for this.

Mary stood and saw her disciples from the room; they eagerly departed to the games room where they met with their families. First Lady Rocc had a bit of shock when she saw her mother and father with her husband and children. Six other sets of grandparents had joined the group and one grandmother, as Bishop Meadow's father had died a few years previous. After saying hello, it was explained they had been brought here to help look after the children as the parents would be away from home and the fathers doing their own work. They had brought only a few items and packed a single case each before leaving their homes and they were told the remainder of their possessions would be brought to the mansion later. Another meal had been laid on for them and now they were all talking quickly and catching up on the day's events.

Half an hour later, a much happier group of disciples entered the lounge to meet with Mary which would change their outlook on the church.

Mary met them with a covered board at the head of the altered seating arrangements. She smiled warmly as each took a seat and looked in her direction.

"I was hoping this talk would not have occurred for at least two weeks, this would have given you time to see how and in what direction I was going. Where mistakes and errors had been made before and why they had to be rectified. However, as we have to see the King and the Pope tomorrow with other world leaders, and right now a number of other world leaders are requesting to be added to the list and Heathrow Airport is almost at lockdown with the number of special planes coming in carrying world leaders, things will have to move more swiftly. Despite my pleas for peace, they still wish to take their armed security officers and that will change tomorrow when we arrive, I will not have guns in my presence, but it is better that you at least know what will eventually happen."

"Is there going to be a change to the world?" First Lady Rocc asked in a very serious tone.

"Somewhat, yes; there will be a radical change in the way you all pray and conduct yourselves. As I mentioned earlier, the Church of England, thought to be at least one of the highest of the churches, together with the Catholic and Anglican versions, God does not wish you to worship Lucifer like you have been doing. As I mentioned earlier, here is the order of the new church." Mary pulled back the large cloth on the board, revealing a diagram of the order of the New Church.

THE NEW CHURCH

GOD

MARY

FIRST LADY ROCC

2nd LADY KING   1st MOTHER MEADOW

FIRST FATHER LANGLEY

2nd FATHER HUNTLEY   3rd FATHER MATHEWS

MOTHER OF THE CHURCH PETERS

Single Women	Married Women
Sister Superior	Mother Superior
Sister Senior	Mother Senior
Sister of Churches	Mother of Churches
Below This	
Sister of the Parish	Mother of the Parish
Single Men	**Married Men**
Son of Masters	Father of Masters
Master Superior	Father Superior
Master Senior	Father Senior
Master of Ceremonies	Father of Ceremonies
Master of the Parish	Father of the Parish

"The disciples are all the same level and under First Lady Rocc or myself; any disciple, will in any church, be the highest person there, unless First Lady Rocc or I am in attendance. As all the high posts have been taken by disciples this should cause no concern to anyone.

"There will be a minimum of two head disciples on any spaceship, which from this day forward will be called ARKS, travelling out of the solar system. A minimum of two disciples shall be left on any planet; these disciples can be new disciples from the people on the planet. Within one month a relay station must be located in close proximity to any new planet visited. Each ship will carry ten relay stations and these will have to be constructed and powered up, then manned by people from the planet you visit. People from Earth can also be picked if they so wish to help control these relay stations and the ships I mentioned earlier will take these people out to the relay stations when they go to collect trade negotiators from the planet back to Earth. They will not be there all the time and will be part of a new organisation called the Space Patrol, they will have a uniform and be able to feel they are part of an organisation that will help people on planets or stranded in space, as they will have access to a number of small ships to go out help those in need.

"The New Church will only use the new bible and hymn books which have been adapted to sing praises to the true God. There will be no further praising to Lucifer; from this day forward, what people here call the devil or horned beast, will no longer be accepted, to give Him His full respect as a true God, which He is, He will be referred to as Lucifer. This is His name and as a God must not be scorned in any way.

"The main day of rest will be on the Sabbath being a Sunday. God will allow anyone to work on the Sabbath if it is for the good of others and to help people in their recreation, this includes shopping at Malls. There will be no fasting and no missing out on any food; all God asks of everyone on the Sabbath, is that they pray for ten minutes; that is all. It can be in their home, at work, or in the park with others, the bigger the group the better to help spread the word of God.

"As you can see, I have outlined the New Church and its ministry, women are in charge of the church, they are the head of the church because it is rightfully theirs; I hope this does not offend you men," she said smiling and looked at the male disciples before her.

"We have learnt a great deal in the last few days with the revelations Master Mathews brought us. We have come to expect the unexpected and will of course follow your commands as you preach them. May I ask when you will be introducing your New Church to the rest of the world," First Father Langley asked.

"My Mother and I have given this a great deal of thought, as I explained earlier, it was hoped we would have time to talk to all church minsters and gradually ease the New Church in; alas, with tomorrow looming upon us, this will drastically change things."

"I take it you know what is about to be asked of you?" First Lady Rocc asked.

She had a good idea herself, knowing what a deceptive government there was in the United Kingdom and other governments around the world. They would lie and cheat and change the rules to suit themselves. With the King now tossed out of his seat of the church, they would be in turmoil. All hell would be breaking loose; she imagined the arguments going on inside Buckingham Palace as they spoke.

"I have a very good idea of what is about to be asked of me; I will not tell you, but I know you all guess governments are very slow to bring about change, especially when they are set in their ways and have managed to corrupt the people of the land for as long as your government has. They even made Queen Elizabeth II abdicate in favour of William after Charles was declined due to his divorce and the church did not want to appear bad in front of

God."

"I think most of the country knows this, many never wished Elizabeth to abdicate and I think this brought on her untimely death five years ago, They assumed the second coming would happen much sooner but were obviously very wrong," Mother Meadow stated.

The others nodded in approval but they could see the look of worry on Mary's face, she was obviously very concerned about the following day. She had not been on Earth a full day yet, now the government, heads of the church and other world leaders, were hounding her like they had Jesus in his time. This time however, no matter what was said, what they were subjected too, they would not let Mary down. They had all decided together as a group and contacted their other disciples by telepathy, that they would stand as one right behind Mary, whether she knew this or not, they had no idea, but they would support her.

"Like I said, I was hoping to have time with you, to get to know you all on a more personal level of friendship before you had to take up your posts. As from tomorrow morning at Buckingham Palace; we may as well start as we mean to move on."

"As of now, providing you all you agree," Mary paused and looked into the faces of her disciples; going deep into their minds; reading their innermost thoughts, to ensure there was no mistake, shame or fear; or if they were unsure they would let themselves and her down so her project would be doomed from the start.

What she found surprised her, not one person was frightened of being shown up against their faith. They all believed in God and Mary and were all fired up, ready and waiting to protect Mary and tell the world about God. Not only that, Mary was pleased to see that behind her back, they had got together and spoke to all their other disciples and they too were right behind her, ready to stand as one against the government and the world leaders. They were also looking forward to the New Church, forgetting already the teachings of the past and bringing new members into the New Church of God.

Mary smiled and held her hands out before her. "I could not have picked better disciples myself," she sighed. "As of this moment you will all assume your new posts in my Mother's New Church, providing of course, you wish to remain in the New Church as My Disciples."

She waited, already knowing their answers, as they all nodded in approval and faced her saying they agreed.

Mary breathed a huge sigh of relief; her message was through to her back bone of disciples. Later in the night, when they were all asleep, she would send a message to the other disciples, informing them of the New Church, reading their thoughts on the situation, something they were not expecting. If everyone agreed, they would be told their positions in the New Church.

"If there are no more questions, I suggest we all retire; tomorrow will be a long day and there will be much to do."

Everyone rose in silence and kissed Mary's hand on leaving the room. As they returned to their own rooms, they looked forward to the following day, to be in their new positions in a church God had prepared for them. They would be spreading the word of this New Church and the True God to people throughout the world and eventually, the Cosmos.

CHAPTER 17

The following morning, they departed from the house at 9.30, dressed in the clothes of the New Church. A dark red skirt and jacket, white blouse, tan coloured tights and red matching shoes. Each woman carried a small red handbag which had a long shoulder strap to hold their bible, a single credit card in case they need to purchase anything and a small version of the new tablet which had been made overnight in Heaven so they could take notes by just thinking at the tablet or talking to it. On their shoulders was a small strap so they could clip their bag strap onto their shoulder ensuring both hands were free to greet parishioners.

They also wore a violet sash beneath their jacket, wrapped around their waist to show their position in the new church. Each person was given a new gold cross and chain made in Heaven which was worn around their neck and was on show to the public; they also wore a pure gold wedding ring on the index finger of their right hand to show they where a disciple of God; once on it could not be removed except by God. The only difference for the men, they wore the same colour jacket and trousers, with the same colour red shoes and socks. They also carried a small red bad which held their bible and tablet and credit card.

Makeup was allowed, so too other jewellery and they were supplied with a warm over jacket for the severe cold conditions, again in dark red. They also work a long cloak, again in deep red with gold embroidery showing the New Church and its shape. A badge of the new church was also handed out to the disciples just before they left by Mary. On their coats and jackets, their position in the New Church was embroidered in gold thread. Mary wore the exact same clothes as the women so nobody looked out of place and for today, they all would wear their long cloaks.

They arrived on the moving coach just as they entered London, the road ahead was cleared of traffic by the police, although these were no ordinary police officers, they were from Heaven and riding the motorbikes they had always wanted to ride during their Earthly lives.

Very soon they entered the gates of Buckingham Palace and the police escort stopped outside the main entrance. There were thousands of people waiting to get a glimpse of Mary and her new disciples and they were not disappointed.

When the coach stopped and the group departed, Mary and her disciples walked around to the front of the coach and stood still for the press and photographers. Among them, in his new red uniform was Mike Adams, who gave Mary a thumbs' up and sent her a message.

"Thanks for the uniform, it's great," he said using telepathy.

He had already written his notes beforehand, following exactly what Mary had asked of him to pass onto the press. When Mary and her disciples waved at the onlookers, they all erupted in applause and then to Mary's delight, they all said the Lord's; Prayer, which had been changed overnight to Mary's prayer, which everyone appeared to know by heart.

The voices joined together and boomed out the words so even the people inside Buckingham Palace could hear the throng of voices and they all sounded very happy to see them. At the end of the prayer, everyone cheered and remained in their positions, hoping to see them appear on the balcony, but many thought the King would never allow it. Once inside the Palace, they found they did not have to wait, they were shown to a great hall which was filled with politicians, church minsters and world leaders. With them were the Pope, the King and Queen of England, the Netherlands and two Kings from Africa, there was also the Presidents of America, France, Canada, the Governor of Australia and many of the people of high order who represented countries around the world.

Just before they entered the hall, a palace official approached Mary. He bowed his head and knelt before her.

"Your holiness, I hope this is the correct manner in which to address you; I have been told to ask you if when you enter the hall, you will walk slowly and bow to the King and Queen of England and the other Kings and Queens and heads of State gathered behind the doors. I have tried to explain, that they should be bowing to you, but they will not have it. I also know this is not what King William and Queen Kate wish to happen. They have suggested to the Prime Minister that they should all kneel to you, but the Prime Minister is adamant and has even gone so far as to threaten them with holding back payments to the crown if they disobey his orders."

"Please stand Anthony, thank you for trying to help the imbeciles who think more of themselves than of the daughter of God. We will follow you in and they will bow to me, have no fear, for I am with you. I see you have a new faith and follow the true God."

"Of course Mary, I have always been a great believer in God, and to hear we had it all wrong does not deter my allegiance to the true God, your Mother."

Mary laid her hands on his head and for a moment he glowed, then bowed his head low.

"Go forth my son and help spread the word of God; after this day, you will join Mother's disciples who will come for you and will pass into the palace to tell you all you need to know. Now Anthony, please lead the way into the Lion's Den," Mary said with a smile.

"Lion's is not becoming of them they are more like Hyenas," Anthony replied honestly. He turned, smiling and feeling much better than ever before, with much stronger legs, as he had previously walked with a limp and suffered with stomach cramps for years. Now he was better, renewed, ready to go out and fight for that which was right and true.

The giant doors to the hall were opened by aids on the other side and the group of disciples, led by Mary with First Lady Rocc and First Father Langley behind her; entered the hall. Everyone was shocked, as Archbishop Langley had not been seen or heard from since the change to Westminster Abbey, many thought he had been killed, but here he was in the flesh and representing the new church.

As they passed through the crowd of people who had all stood when they entered, they shook hands with Mary and the disciples, everyone bowing first like Mary had hoped they would without any pushing. It took them ten minutes to get to the top of the hall where the King and Queen of England were seated on their thrones and beside them were the President of the United States and President of Russia. The UK Prime Minister with his other ministers and leaders of the opposition were all in attendance waiting to have their say.

Everyone, including the Pope had kissed Mary's hand and spoke with glad tidings and wishes of hope for the future. As they approached the King and Queen they remained seated, and it was President Yuri Dmitriev who stepped forward, bowed, took Mary's hand and kissed it while on one knee.

"It is a great honour to meet you Mary, may you have a very long life on our planet and prosper greatly," he said then stood, smiling at Mary.

"Thank you President Dmitriev." Once again she held his hands tightly and after a few moments he felt like a new much younger man, ready to help this woman if she needed it.

As Mary turned to face King William, she saw Queen Kate stand and then taking a large swallow her husband followed her lead and they bowed their heads. They then each kissed her hand and spoke quietly with her. It was obvious the Prime Minister was not amused and thought something was wrong as this was not what he had planned to do in throwing Mary out of their county and having her house demolished due to no planning permission from the local council.

"It is a great privilege to meet you again your Holiness," King William said. "Please everyone, be seated, you too disciples of Mary he said with his hand sweeping to the empty seats that were ready for them.

Mary was surprised and so too were the disciples; everyone waited until the group were seated before King William and Queen Kate took their seats. As Mary looked around, she noticed the palace aid who had spoke to them earlier was now trying to get her attention and rubbing his head. Immediately Mary got the idea and read his thoughts. A moment later she spoke to the driver of the coach and her special escort. He closed the coach doors and drove out of the palace, disappearing into the heavy traffic. Despite a heavy police search, they could not detect the coach on any road leaving London; as it was now parked outside the Mansion of Disciples.

"Mary, we have asked you here today to discover what it is you intend to do while you are here. We have noticed your mansion, but several people have objected to its size and situation on prime building land, there is also the fact that there appears to be no planning permission for this building and time given for people and authorities to decide if it were in the best place. For the size of your building, we could have placed a thousand houses there," King William said, and glanced at his Prime minister who nodded in approval and smiled at him, thinking they were back on line.

"I see," Mary replied. "I am afraid I had nothing to do with the building of the mansion or the position it was placed. It is as you say in a prime location and has lovely beauty spots around it. Neither did I have anything to do with the rebuilding of every derelict house in this county or all the houses which have been added to it for the benefit of the homeless and poor. God did all of this, it was not of my doing; She alone is responsible for this as too She is responsible for all the healing that has been done and curing all the cancers and AIDS. She has stopped everyone smoking, taking drugs, and drinking too heavily, which She has not banned altogether. The weight problems people had are gone through my Mother."

"But you fail to see the position you have put this government in," said Mr. Templeton, the Prime Minister for the United Kingdom, who had stood and walked over to stand behind his King.

"Tell me, has God not saved this country millions in NHS funds, solved all your drug problems and put things to right? Has She not stopped the malnutrition around the world, stopped all the homeless people around the world, emptied hospital beds and cured people, surely that is what your doctors were trying to do anyway," Mary replied.

"You seem to have missed the point, the government collect taxes upon many of these items, tobacco and alcohol alone gives us a great deal of money which we in turn give back to the people in other forms of assistance. You should have discussed this with us before making sweeping changes which have affected the government's handling of the country. We need a great deal of the income to pay our wages and keep our buildings up together and keep us in the state we wish to be. I know we work long hours and like a drink, but we move slowly and this is done deliberately so that we as a government can be seen by our people to see we are doing things in the correct manner. Not only that but we have just given ourselves a twenty per cent pay rise, where will we now find the money for that if people don't buy alcohol and cigarettes?" He suddenly realised what he had said and was thankful the people and unions didn't hear it and blushed profusely.

"What we are saying . . . or I am saying; to be very blunt about it; is we wanted homeless people. We wanted people to die from cancer and AIDS, tobacco and alcohol related diseases and such the like. We make money from the drug Barons, people make money from drugs; we employ people to catch the drug runners and enjoy putting them in prison, even if again it costs us money to look after them in prison; it is a way of life for millions of people.

"After all, what will the police do now, the prison wardens, you have released and cured all the prisoners, people will no longer commit crime because they will not need to. They have enough food to feed themselves so don't have to steal it from the shops. The shop keepers will no longer need to employ or install expensive CCTV which we, the government get money from in taxes. The more money companies spend on CCTV, the more money we get in taxes to pay our wages and luxuries. It is costing a fortune, over three million pounds to have new more comfortable seats put in the House of Commons, we want more computer games and I have personally just ordered thirty wall screens at twenty thousand pounds each for the government homes in Downing Street which have to be paid for. We also need money for weapons, ships, aircraft, to pay our armies, rockets to blast the hell out of the Ruskies, the communists must be halted before they overthrow the world and we all live in their undesired way.

"I would rather it be like before you arrived, at least then we would survive in our jobs a little longer with the people hoping we would do something; protecting them from the world of evil and oppression."

He found he was saying what he shouldn't be saying, but he couldn't stop himself, it was impossible to tell a lie, he could only speak the truth. His ministers, like everyone else in the room and especially President Yuri Dmitriev, looked at him with horror filled faces. This was not his planned speech, what they wondered, was he doing?

"I am Mr Barnes, minister of Defence, do you realise what you have done? If this planet is at peace because of your Mother's intervention, what are we going to do with all the people we employ? All the money we have planned to spend on nuclear rockets, whether we use them or not, this would advance us in nuclear research to far greater prospects. All this nonsense about travelling into space, other beings, being totally honest," he tried to stop himself but he couldn't.

"Please continue Mr Barnes, we are all very interested in what you have to say, I am sure you will enlighten us all, including their majesties here," Mary said calmly.

"Well, to see UFO's, other civilisations, we've known about them for years, they are nothing new to us, but they are a threat to us: we've captured over ten beings, entities we call them. They're not like us and we couldn't communicate with them very well; in the end they had to be destroyed before they brought thousands of their kind to this planet. Do you know it is our belief; the Government's that is, that they intend to populate our planet and take control? We need better weapons to destroy these beings, not visit their bloody planets and talk to them about God. Let's blow the fucking species off their planets before they come here and take our planet from us. Talk about peace, Christ, their bloody hideous creatures; with no more rights than the animals on this planet.

"This is why we have to put a stop to all space research; of course we are still researching into different space drives, what's the sense of spending countless millions when it will do nothing for us. What we need is interstellar drive; now if you can give us that, then we'll wipe the bastards out of existence with the weapons were are currently designing in this country and America. Do you know . . .?"

Again he tried hard to stop himself, what he was about to say would ruin all their plans; only those people in high places were supposed to know the truth. That did not include the King and Queen of England or the Prime Minister, certainly not members of the church. They loved their people; they wanted a good relationship with them, not war.

"The thing is, we captured one of their ships, and the Americans have two ships and ten creatures on ice. They dissect one a year to discover as much as possible about them; the Americans still haven't broke into their ships but we have. We discovered all sorts of things; there is a new form of laser, so powerful it will knock a satellite from the sky from where we stand. We haven't got to the engine compartment yet, but we will any day now. When we do we'll strip apart the engines and learn how to build our own and then we'll beat everyone else to the stars. With lasers and other things that ship carries, we'll rule the world and then the universe."

He stopped and looked at Mary, then the PM and his sovereigns; he knew he shouldn't have mentioned those things, but he couldn't stop himself. Something made him talk, continue to talk and tell everyone who shouldn't know the truth, the whole truth.

"Mr Templeton," Mary said. "You have people within your government and civil service who deceive even you. It is about time you have full knowledge of what some of your trusted people who advise you do," she continued.

"No!" Mr Barnes shouted at the top of his voice, standing instantly. The guards behind raised their weapons; as it was an order which was already prearranged for them.

"Not you lot, back to as you were," he commanded and turned bright red, everyone had now seen through his plan.

"The thing is we cannot allow the new people to infiltrate our already overcrowded religion." He turned to face his peers, the cabinet, other ministers and leaders.

"This women must go; just like her brother before her. If she is allowed to continue to spread the word of God as she wishes, freely unchecked, it will throw the entire planet into chaos, we were far better off with homeless people, those dying of cancer and Aids, I have it all arranged; it will appear Mary tried to take control of our minds and killed everyone. I will naturally take over as Prime Minister and then become King Consort to Master Prince George," he admitted, still shouting his plans to everyone.

"Don't you all see; we need total control, to do as we please, to run this country as we wish to?" he continued.

"Guards," called King William, "Put your weapons on the floor and file out to your commander and take him with you and place yourselves under house arrest until this can be sorted out."

"There is no need to concern yourself with their guns your Majesty, they would never have fired," Mary interrupted. "I disarmed them while we were here, as I have disarmed every special agent's weapons they have on them. I would never have allowed gunfire in this building. I'm afraid it is Mr Barnes who needs to be questioned further with your Prime Minster. However, for your own piece of mind, when we have completed these discussions, I will allow you for one week, to read all your cabinets and civil servants mind's. I will leave it up to you to take out the men and systems which are being used for unfair treatment on your people.

"As for the aliens in this country and America, I have returned them, together with their ships to their home planets. In future, space research will be approved through a board I will set up, it will be operated by disciples as these people can read minds and ensure no dangerous tasks are undertaken in space. Now; I would like to get down to the real reason while my disciples and I are here. You may sit down Mr Barnes, I have work for you later," Mary said firmly. "Guards, secret service agents, you may also leave us and get yourselves coffee or tea and refreshments while you are waiting," Mary said firmly and then placed a large table of refreshments for everyone to help themselves in the hall as she talked.

They were amazed at her miracles and continued to listen without further interruption. It took two hours to explain in detail where man had gone wrong with religion and what God now wished to happen. She ended up with her description of the New Church and said it was now formed and introduced its backbone and leaders.

"Now I have explained everything about the church, I would obviously like everyone to change and conform to it. In short, I would like an amalgamation of all the churches into one church, as it will make everyone much stronger," Mary said with a smile.

"We are all very fascinated with what you have said, but this surely means that what we have all believed in for two thousand years will come to an end," Mr Templeton said in an angry tone.

"All good things come to an end at some time; your reign as Prime Minister is at that point right now," First Lady Rocc said. Mary looked at her astonished that she would intervene in her task, but saw the point.

"As you put it that way, I suppose they do. I think in the back of every man's mind he knew and recognised that women were superior to them. The only way women could be kept down, was to keep them in their homes and make their homes as comfortable as possible so they would not have to go out to work. As they got into business and Mrs. Thatcher remained Prime Minster for over eleven years; she showed women they could not only do a man's job, but run a house and often two houses and look after children as well, begging your pardon Sir," Mr Templeton said turning quickly to face King William.

"As head of the current head of the Church of England, I will give my bishops permission to amalgamate with the New Church. Speaking for myself, I am willing to subject myself to the new order of religion," King Williams stated. He looked at his wife who was smiling, holding his hand. It was for him she knew, a great come down to accept a woman as senior to him.

"We must all do this," said the Pope. "I have been misled and interpreted the bible incorrectly. I have sinned greatly against the one God I seek to serve. As head of the Catholic Church, I will inform my subjects from the balcony of this palace, if I am allowed, of the immediate changes and I decree I will accept the amalgamation of all churches into one true church.

"I can assure you it will cause problems at first, but once the people understand what is actually happening and what errors we have made in the past, some of which I am truly deeply sorry for, they will see the light and turn to the real God," said Pope Paul.

"I think you could use our balcony for such a statement and Mary, I'm sure the people of London would love to see you as well. Will you accompany us later when we have completed our talks here for a few minutes to face your people?" Queen Kate asked.

"Of course," The Pope and Mary replied in unison.

With the other assembled leaders of the churches, the agreement took place but all the time, Mary and her disciples read the minds of the politicians. Mary could see many politicians were against the new church, despite what their leaders were saying. Even the Prime Minister was having reservations about this new form of religion. There was some doubt in his mind, a question to be answered,

Mary had shown him deceit was everywhere and he was now trusting no one; including Mary and her disciples. He was unsure that other leaders would have the same opinion, in the past, and now, they were always at each other's throats. War was sure to break out at some point, no matter what; he had to have the troops and the weapons ready to fight. He could not rely on Mary alone to set the world's politics to right; it would take time, much longer than a few hours at Buckingham Palace.

Just because King William agreed, it did not mean the Russians or far eastern leaders would; their beliefs in God were very different to theirs and of course, to what Mary was proposing. In some countries, women would never be held higher or even equal to a man. Man was the person in charge; women the scum of the Earth, and nothing more than to look after children and bear their men's offspring.

"You will have your work cut out in some countries I can tell you," Mr Templeton said smiling at Mary, trying to hide his innermost thoughts.

"No doubt I will overcome them; we shall all overcome our fears eventually."

"What about Lucifer, surely He will try again to sabotage your work here on Earth?" Queen Kate said, showing concern for a fellow woman.

"He has not returned yet, I just hope He will see the error of His ways and see the light," She stopped there not wishing to elaborate on her remark.

"What about the problem with the local councils with regard to the housing and extra houses that God has given us?"Mr Templeton asked seriously.

"You mean to tell me, when God does help you, by doing more for you in the past three days as than Lucifer has done all your lives, you wish Her to return everything, including the health of people to their former states? I definitely would not like to be in your shoes when the press hear this news. Your party would be kicked out of office this very day I assure you. The people will lose faith in you and demand your resignation immediately. Have you really thought this through before I ask God to put it all back to as it was? She will if you wish, but all those people who were ill and have been cured, will know it was you who put them back to what they were. Mother will not wipe their minds clear." She paused for a considerable time.

"Of course, if you wish it, then it shall be done, however, before you make your decision, the press who are outside this building are hearing every word you say, everything that has been said here today, has been relayed to the press and the people outside the main gates. So, what do you wish me to do?" Mary asked slowly.

The Prime Minister looked at his aides who were shaking in their shoes. They had all turned white with fear, the consequences of all the housed people, the well people, the blind who could now see and the deaf who could now hear, the dying who were cured. Then there were the smokers and drinkers, drug addicts, the ones which brought in the real money. The weapons which cost billions; millions of this money was taken from other budgets to spend on missiles and fund the services elaborate uniforms laden in Gold Thread.

Making up his mind, Mr Templeton did not wish his career to end in a lynching for the sake of local councils, they would have to sort out their own problems, and he wished them well with that, knowing they would never win.

"Please, stop," he shouted, raising a hand to Mary in a vain attempt to put things right, "I will sort out the councils myself, do not alter anything which has been done or you are about to do I beg you, the people will benefit I agree."

His ministers looked at him, shaking their heads in disapproval, the people put them in office and they would support them in their hour of need. Mary looked at them all with contempt; she could see the greed in their eyes and in their thoughts. All they wanted was to line their pockets, never mind the sick, elderly and disabled; as long as they did not fall into those categories they didn't care.

She smiled, looking about her. "You fear so much and need to; you have enemies within your cabinet Mr. Templeton. When all of you die, God will wish to discuss your life with you and what you have done to assist people in your elevated positions."

"But what about the planning permission for your mansion; some people and individual companies who could have made a great deal of money from the land are outraged that God, or you, should take these lands from them? What do you intend to do about it when people are not asked, but told by councils and courts of this land, to pull down their houses you or God have built?" Mr. Morris, the Minister of Housing asked in an agitated voice.

Mary glared deep into his eyes, beyond into his deepest thoughts. He turned red as he felt Mary dig deep inside his head, uncovering secrets which he wished no one to know.

"Mr. Morris, you hold office in this government, you could have made lands available for housing years ago, yet you kept it for your own desires to make money. You have secretly built up ten companies who you direct, to purchase land which you give out and make a handsome profit. You intend to make over 40 million pounds from this land deal, is that honesty of a man in your position?" she asked him.

His cheeks glowed bright red and he remained in his seat.

"You also have a wife and a mistress, two children by your mistress, yet you do not keep them; that is despite the strongest laws this country has about such matters. You forced your mistress to lie for you so you could remain in office; it is about time you resigned from your position and faced the consequences. Don't worry, the prisons are not operating now, but you will be forced to accept your children and pay for them and with the hundred and eighty five million you have in your sixteen bank accounts, you will give your mistress and your children the money they deserve and your name," Mary insisted.

He looked around at everyone who stared at him in disbelief, the last person his eyes met were those of his wife Marie and his Mistress Jeanette, with her two children Sean and Timothy, eight and nine respectively.

"This is outrageous, how dare you delve into my mind and destroy my life?" he bellowed, changing the subject as fast as possible from the land.

"I will ensure the houses are crushed to the ground through the courts of this land!"

"Mr. Morris, I will ensure you pay for your mistakes and you will bring up your children correctly. You will pay for them but not through a government office. You hereby, by the authority invested in me as your King, are dismissed from your seat of office. You will never be allowed back into politics again, by my law!" King William said very sternly.

"You can't do that, you can't sack me, only the PM can do that," he turned to look at his Prime Minister for help.

He shook his head from side to side, looking very sad indeed. "His majesty has that power, if he did not dismiss you, I would have and I will at this very moment, so you have been dismissed twice."

"But the land, the houses!" he screamed and finally broke down in tears.

As he went to stand he fell to his knees, he tried to breathe but found it difficult; his lungs were labouring; his back started to ach and he looked up to the people around him. Suddenly, a wheelchair appeared before him and around his face, two clear oxygen pipes connected to a small oxygen cylinder. Mary looked at him and he was lifted from the floor and placed into the wheelchair which was self propelling, meaning he had to push the wheels with both hands to move it. The heavy oxygen bottle settled itself into the bag at the back of the wheelchair. His eyes were blurring and he wondered what was happening to him, for a moment he thought he was dying, then realised it was all Mary's doing. He would have to cope with this for a few minutes as she was obviously testing him, making him repent for his miss deeds.

"Okay Mary, I apologise for everything I have done and said. I will do as you wish, now please, take this away and allow me to continue my life looking after my two sons," he said through gasping breaths.

"Mr. Morris, Henry, you have done wrong, you smote the elderly, disabled and blind, wishing them to suffer. I did not do this to you, but I did put you in the chair as I could see you were not capable of doing so yourself. You have advanced emphysema and spina bifida in its mid stages; you can no longer feel your legs and will never walk again. You will from now on require oxygen to breathe, without it you will die. I suggest you attend a hospital as soon as possible to get diagnosed. Oh thank you Mother." A piece of paper appeared in Mary's hand which she handed to Henry.

"This will tell the doctors at the hospital what is wrong with you and the medications you will need for the rest of life. They will set you up with oxygen around your house and oxygen bottles to take out with you."

Henry took the paper and tried to read it but found it impossible with his failing eye sight. "I can't read this, my eyes, something is wrong with my eyes," he cried.

"You will have to see an ophthalmologist as soon as possible; he will set you up with glasses. Oh, by the way, you will not be able to drive from now on either; if I recall you live in an apartment with five flights of stairs, I suggest you buy a bungalow as soon as possible and for now, move to a hotel on the ground floor," Mary suggested.

"How long will this last?" Henry asked sounding very frightened.

"Until you die, which will not be for the next fifty years at least. You wanted My Mother to turn back time, She is now showing you what others were having to go through, enjoy your life Henry, *live long and prosper*, as Spock used say," Mary said holding her hand up spreading her fingers in the Vulcan way.

"But, please, help me," he said to his wife then his mistress.

Both turned their backs on him and walked from the room. He turned to look at Mary and then he was gone, he materialised outside the room next to his wife still in his wheelchair.

"Will God leave him like that?" Mr. Templeton asked.

"For a week or so, until he has repented and sees for himself what people had to go through. It will be his longest week ever I assure you, and very embarrassing for him," Mary replied.

"To bring this matter of planning permission to an end, planning permission for all the houses, my mansions and the alteration of every church in the land, Malls, factories and other buildings which have been erected or repaired has been passed through God's eyes, all the relevant paperwork and documents are now in the council departments which require them. Every plan has been passed and nobody has objected to any building," Mary said finally.

The politicians gave a great sigh of relief with the problem solved and then, a newspaper appeared in every politician's hand which they unfolded and read the headlines.

Prime Minister sold out help from GOD! All for the sake of Red Tape!

They all read the story which followed which went on for five pages naming everyone in the ministry and what had happened to Henry Morris.

"I fear we shall face a reprimand for this in the house tomorrow," Mr. Templeton said.

"A lot can happen when you deal with my disciples or me, things do have a tendency to move quickly. It is best to make the correct decisions first, your career may depend on it," Mary explained.

The talks continued for a further two hours with all the people being made members of the New Church except a few members of the cabinet and Mr. Templeton, who decided to wait for a while as he had never been a big believer in any faith. After a short wave to the public from the famous palace balcony, and the Pope telling the people around the world he would bring his church to the New Holy Church, the group of disciples and Mary disappeared from the palace and materialised back at the Mansion.

The other disciples which had been made overnight and during the previous day were assembled in the main hall. Even First Lady Rocc was surprised at the number of people which Mary had involved in her project.

They entered the hall to applause from the two hundred and seventy five new disciples gathered before them; it was obvious the group had seen what had happened at the palace.

"My disciples, I am happy to tell you that the first stage of our work has been completed. Those who I asked to read the minds of ministers attending the meeting, I would like meet with later, unless there is anything immediate I should be informed about," Mary said in a very happy mood.

Everyone was silent except one woman, Gemma Kennedy, who was sat at the back of the room. She had been silent even when Mary arrived; she did not stand or applaud their work, but appeared to be in a trance of some form. She was somewhere else and she looked very upset and concerned.

"Mary," said John, who was sat by Gemma, and brought the entire room to silence."I think something is wrong with Gemma."

Mary walked up to her and knew instantly something was drastically wrong with Gemma. Placing her hands on Gemma's head, she saw and heard what Gemma was witnessing, this in turn was passed through to the main screen on the far wall.

"You all must see what is happening surely?" Mr. Templeton was saying to the group of ministers at Buckingham Palace. "Mary is going to be a pest to the world, with the assistance of God she will ruin everything we have set up. The progress of man may well be halted because of the lack of war."

"That's ludicrous," Pope Paul replied in a very angry tone. "We have been struggling for world peace for hundreds if not thousands of years and now we have it on our doorstep and God, not some minister or president, God has decided to give us world peace. You just cannot go against God, and let me remind you, unless you wish to join Lucifer when you die, then you will have to face God, the very God you are smiting right now. With God's help, we will go deep into space and take our people out there to meet other extra terrestrial beings. Surely this is what you want as well?"

"Yes, but without weapons; God only knows what we'll discover; and I mean that literally as God will be telling us where we will go. We need to be prepared for anything, we don't know what type of beings there are out there; we need to be able to defend ourselves, kill if necessary or possibly be killed."

"You always want to kill people and other beings," said King William interrupting "I do not wish to have war, peace will be better, to live in peace and harmony with our neighbours, what a world this could become. Mr Templeton, you are nothing more than a war monger, I remind that I have been on the front line, fighting in Afghanistan and so too was my brother Harry. We both saw our friends dismembered and killed, all you have done is send soldiers to war; you haven't been there yourself; it's not a pretty sight I can assure you," he continued raising his voice.

"That may be so, but I'm in power and will remain so until elected again in two years time. This is if I do not manage to push you Royals out before then. We are on the verge of a Democratic State, soon you will have no say in the matter, you will be no more than normal people looking for a job and claiming dole money, except for you your Majesty, you'll be sent back to the air force and Harry can continue in the army, for when I'm in power, you'll get nothing from the government and I will be residing here, not you. I'll take away your houses and land if I can find a way," Mr Templeton said joyfully.

"Mr Templeton is the most disgusting man I have ever met; I thought our Prime Minister was made of much better stuff than this. I have read him and he is not controlled by Lucifer, it is bread into him, war is in his thoughts day and night, he wants to keep the population low, at least in the third world. It is a place for the latest designs in weapons to be tested, at the cost of human life; he is not worried who lives or dies, as long as this country gets money, which is also lining his pockets," Gemma said, giving everyone the rundown of the Prime Minister.

"I will give you all my promise; I will stop, by hook or crook, the building of those so called space ships. I will oppose every female job there is and put the women back to where they belong, in this country at least. Every person who is currently working on the space project, who is not attached to normal space programmes, will be thrown into prison and executed for High Treason against the crown," Mr Templeton said banging his fist onto the table before him.

No one was happy in the palace, everyone including most of ministers were siding with the King and Queen and members of the church.

"It is obvious the Prime Minister will try to get some people on his side and he can influence a great many people. I will speak to my Mother about this as it has far greater implications than I first thought we would be up against.

"However, we will not stop our plans from going head. I am sure the Pope, as leader of the Roman New Church will spread the word immediately when he returns in two days to Rome," Mary explained.

"That is if he returns," said Disciple Andrew, speaking up. "Mr Templeton is so furious, he may try to kill the Pope and blame us in order to gain backing," Andrew continued.

"You are correct of course, I have a feeling the Prime Minister of the United Kingdom, may not be the only leader we will come up against who opposes us. Gemma, I will change your tasks and give you a few extra gifts; you are now assigned to watch over the Pope, or Father of the New Church of Rome. You will travel with him and I am at the moment informing Father Paul what is happening. You will depart after this meeting when I have had a further word with you. I will expect you to keep a close eye on him whilst you are still here with us.

"Disciples Andrew, Joanne and Philip, I would like you to attach yourselves to the other high church ministers and stay with them until they have safely returned to their countries. The rest of us will see as many church leaders as possible within the next two weeks; please be on your guard for any trouble from politicians and warn others if there is trouble and do what you have to, to save people in danger. Do not forget, the police cannot harm you or detain you, you can leave their cells whenever you like, they cannot kill you, they cannot break your bones or hit you, if they do, they will find they are with my Father in Hell and they will not enjoy it there.

"I have decided; so as not to complicate things, the High Disciples will be those who are High Members of the Church. If needed, you will report to them anything which is needed and they will deal with it immediately. I did not realise I would meet so much opposition so fast, I had hoped in your era, you would give me a chance without being put down before the world had heard about me and my Mother . . . the True God. The real truth which has been hidden for endless years. I would like to talk to all individuals right now, and this will only take a moment as I will be speaking directly into your minds."

"Before you do this," First Lady Rocc interrupted. "I would like to say, we will spread your word, the word of truth and justice; the real God will get a chance to see Her children, whether the politicians like it or not." First Lady Rocc held Mary's hands and knelt before her.

Something inside her, women's intuition, told her Mary may stay on the Earth for a very long time. One thing was certain, Mary would not go out as her brother did nailed to a cross, She would just disappear into the night, possibly with her followers close by, but there would be no crucifixion of a woman, Jean Rocc would see to that personally if she needed to. She had come to love, cherish and admire Mary and would follow her to the end of her days.

As each disciple concluded their discussions with Mary, they disappeared from the room, going on their own specific journey to carry out the duties Mary had given them. Of her remaining eight disciples, she spoke to them in a single group.

"I have sent each disciple on a specific mission; that is except those which are on protective duties. I have sent two disciples to each country in Europe; there they will find ten more special disciples, these people will be dedicated and relentless to the cause and have a special gift, so they can protect people. Once these people have been sent here to me, the disciples will spread the word of God and make three hundred disciples in each country and change the faith of those people there. They will find I'm sure, that many of the poorer people will flock to the new churches and believe in the new God," Mary explained.

"If I may ask; who will these new disciples protect?" First Lady Rocc asked.

"The very people who run our churches and will be at the mercy of the governments until they come to see the way I am moving forward. If we do not protect these ministers, then the people will fear coming to church to pray to God; especially if Mr Templeton gets his way and sends armed police, whether the guns work or not, the public will not know the guns don't have ammunition in them when the police burst into the churches. The police could try to arrest people and hold them in huge areas to stop them joining our church saying it is Treason against the King."

"There is also the problem with the spaceships and the people who will make them. Those people too are at risk and will need our protection against our government and other governments around the world," Mary added

"But surely, wouldn't it be much safer to bring all those involved in the projects here, or even create a place for the group and their families to work in outside the Earth like it is here. Somewhere where it would be safe and away from prying eyes of the government?" First father Langley suggested.

Mary looked at him as if probing deep into his thoughts. "First Father Langley, sometimes you pick the most inopportune times of butting in," she said coldly.

Everyone was amazed at her reaction to his good suggestion; but he stood his ground, he didn't even blush but smiled at Mary, who suddenly burst into laughter.

"You are correct of course, to the last detail in your head. I have already told the protecting disciples of your plan, they are at this moment talking to the designers, but which country and where?"

"America is by far the largest country; however, they have the coldest police force by far. Here you would just have to find somewhere to put the spaceyard, but the land you need for building such vast ships, it would be far better in America," First Father Langley suggested.

"You are contemplating another site in your head?" Mary asked almost laughing and quickly, with his permission, read his thoughts.

"Of course, the vast areas of Salisbury Plain, which could be bought by the Ministry of Defence for no specific reason and we could use that, surrounded by an invisible screen which allowed the people to work freely, while on the outside, it would appear as if nothing was happening, just plain empty land, I like the idea very much, so much so that we have just acquired the land in question," Mary concluded.

"If I may make another suggestion Mary, President Yuri Dmitriev seemed to be on your side, in fact I got the impression the Russians were more behind us that the Americans and he did say anything he could do to help us he would. Russia has huge areas of unused land and there is one area of land I was thinking of in particular and that is Chinoval: the nuclear power station has made the entire area a no go zone. If you could cleanse the area with nobody knowing, and ask President Yuri Dmitriev if he would help us, we could have another spaceport there and nobody would be any the wiser. Even with lorries going into the area nobody would know and if we could stop the prying eyes from overhead spy satellites seeing what we are doing, President Yuri Dmitriev would have a good economy when things cool down and go our way again.

"The area would need new roads and an infrastructure, a new town close by with factories to make parts and employ people to do the jobs needed to make the huge space Arks. People would need to move there and we could possible add a new railway system and an airport with planes to ferry people around the area and to other Russian cities. I really think this could work and the Russians do not like the British Prime Minister at the moment after what was said earlier, it would be in his favour to do this," Gemma added.

Everyone looked at her and smiled.

"Yes, I'm sure you're right, I will speak to President Yuri Dmitriev this very moment, I believe he is now at his Embassy in London." Mary paused for two minutes while she gave President Yuri Dmitriev all the information and planted the idea firmly into his head.

"He has agreed to our suggestion and thanks us for thinking of Russia to build the new space Arks. He has said he will personally speak to all his people when he returns to Russia later tomorrow, at the moment, he is speaking to his government using telepathy, which I have given to him and all his government who I felt would use the gift properly. He has asked for assistance to build the infrastructure as it will take far too long to get all the materials ready and the work completed, it would take in his opinion, with twenty thousand workers, at least two years to build the railway, airport and three towns with all the houses and factories needed to supply the spaceport with materials. Therefore, I have just ordered the work be carried out by the angels who are working with us. It will all be completed by this evening, and people will be living in the new homes with all the latest mod cons of America and Britain."

"Could I make a suggestion?" Andrew asked."

"Please, if any of you have any ideas, just tell me," Mary replied.

"What if you could make something happen in Australia that forced the Prime Minister to go there and sort out a problem where it would take him away from the country for at least two weeks and give us time to get things organised. After all, it appears it is Mr Templeton who is trying to change the minds of the people in this country and we might, in that time, get a vote of no confidence in him, and have the government vote in a new Prime Minister who will help the Monarchy and us."

"That is an excellent idea, but what could make him go there?" Mary asked.

"Civil unrest; the people over there demand to support us, forcing the Australian Governor to call the Prime Minister over and sort out the mess. Mr Templeton would need Australia behind him as too the whole Commonwealth if he has plans to depose the Monarchy," Andrew suggested.

"That could well work; I'll do that right away." Three minutes later, while Mary was talking to the Australian Governor, everyone else was trying to come up with ideas.

"It is done, actually, Governor Selcombe is all for us, as are all the people in Australia, he has said we can use anywhere we wish in Australia to build our spaceport and they will help in any way they can. I think this is where we will place the seventh spaceport to build the very large ships that will travel between galaxies. As they have so much land, we could build a second spaceport there to build the smaller ships which will travel from Earth to the planets to bring back Trade Delegates. We will also need a large spaceport in Israel, very close to Concord, something that I had forgot in my haste. I will get my angels to do these jobs and they will tell me when all the work is done," Mary said confidently.

Within a day, Mr Templeton was on his way to Australia to sort out the unrest with an army of secretaries and guards to help him enforce his laws. As soon as he boarded his plane and was out of UK airspace, Mary went to Buckingham Palace and spoke to the King and Queen with their advisors. She then, with their permission, went to 10 Downing Street and spoke to the Deputy Prime Minister Mr Demart, who was behind Mary and against Mr Templeton.

After calling in his ministers and lawyers, they went to work to bring about the change in leadership. Meanwhile, while still in charge, Mr Templeton was trying to get the church leaders to walk into a trap.

Two days after purchasing the land on Salisbury Plain, huge buildings were erected and placed behind an invisible curtain which from the outside looking in, appeared as green land, with barbed wire fences and signs reading '**KEEP OUT MOD PROPERTY**' scattered everywhere.

In Russia, the railway, the most modern in the world, with locomotives and carriages, was now travelling to the new spaceport filled with parts and materials. Four new towns had been built overnight with factories and Malls and decent housing complete with a green power station to give everyone free power, water and gas at their homes and the factories. The spaceport was completed and all the designers and ship builders were now located at the spaceport and had up to date vehicles to get the people to and from work. Again the spaceport was invisible from the air or space and to anyone looking at it, all they could see, unless they were working there, was the barren land of the old nuclear power station.

In Australia, three spaceports with the entire infrastructure had been built and the people were now helping to construct the huge spaceships and smaller ships that would be used in the future. For the benefit of man, there were also another spaceport much smaller, building and testing the new space shuttles that would be used to transport people up and down from the large ships to the planets they visited. The pilots who used to fly the jet fighters were now given the task of learning to the space shuttles. The people were also building a new fleet of fast two woman interceptors that were armed and used to destroy small asteroids or repair the exterior of the ships and help explore the Cosmos. They were also fitted with 3D cameras and telephoto lenses that gave clear pictures of heavenly bodies.

Lasers, against Mary's wishes were added to the large ships, just in case they were needed, and after some thought, decided a few weapons might help the people on the planets move mountains, and create new areas for homes, so with strong guidelines, the armourments were added to the Arks, but not to be used to fire on others.

Another factory was making the droids that would be used outside the ship to carry out repairs and other droids that would be used to go inside the engines to carry out any repairs that could not be done by people. After a further two days, Mary could see her vision for the future was now starting to form.

During this time Mr Templeton's flight to Australia was delayed for a further two days due to a hurricane forcing his plane to land at Johannesburg in South Africa to allow the storm to pass, he was trying his best to communicate with the UK but due to the hurricane, all phone lines were down and water had caused the local power station to close for safety reasons while the storm was going on around them. Mr Templeton was forced to take cover with a thousand other people in an underground shelter, where all he heard was how good Mary was and they were all now praying to God to stop the storm or at least stop it damaging their homes and places of work, as they relied on their work to live.

Mr Templeton saw his chance and used the opportunity to try and put Mary and God down as he was sure if God was all good, She would not put these people through Hell with the storm. He continued for two days to put down God and not once did God stop him, She had a better idea. The storm raged for a third day, giving Mary more time to complete the jobs she needed to do, however, the, traps which Mr Templeton had ordered were now being put in place to lure the heads of churches into.

Father Paul, former Pope Paul, together with his close aids, boarded their plane and started their journey back to Rome ready to start work on the creating the New Church. As the plane's door closed and locked, the press and Mike left the airport. A few minutes later the plane was in the air, flying over the English Channel on its route to Rome. It was dark, a routine night flight that the pilots were used to; what they were not expecting as the captain sipped his coffee at the flight controls, was a bright light appear right over his cockpit. A plane or something like it was almost upon them; he placed his drink in the drink holder to stop it spilling and damaging the controls and took immediate evasive action, trying to avoid a certain collision.

The light was still there, keeping up with his aircraft; then there was second light to their right and another to their left. Father Paul had limited powers; his was the power of multi language and scan reading with a perfect memory and telepathy, he could do nothing to assist himself or the others on the plane.

This however, was not the only plane in trouble; two other aircraft which left within minutes of each other were also having the same strange lights pass over the aircraft. UFO's? They were definitely not civil aircraft, no civil or air force jets would do such a thing to these aircraft the pilots were sure. The pilot even doubted the air force had planes in the air, especially fighters, since Mary had come into their lives, most fighters had been grounded as their weapons would no longer work.

The pilot tried desperately to call ground control, but all he got was static; however, the objects following his plane did show up on his radar as small fighter aircraft. The pilot finally assumed these were small UFO's which were jamming their communications and 'buzzing' them, as it was said among pilots.

At the rear of the plane, Disciple Gemma was looking outside her window with the small overhead light turned off so she could see better. She found the five objects and thought about her eyes, enhancing her vision a hundred fold. They were not alien craft, they were not civil aircraft but fighters of British origin which had been stripped of their weapons and were carrying extra fuel tanks.

At this moment they were spraying the jet with aviation fuel; it was the only weapon the air force had at its disposal, but they were still too high for it to do any damage. It was so cold outside; the fuel was starting to freeze and with their speed, it was swept off the wings and away from danger. Then she realised the engines were on the wings and if the aviation fuel was ignited, the plane would burn and fallout of the sky.

She checked their height; the pilot was still descending at an alarming rate, being forced down by two fighters over the cockpit. She got up from her seat and walked the length of the cabin, placing a firm hand on Father Paul's shoulder as she passed him. Instantly he knew this was no ordinary woman, she was a disciple, and now he knew they were in safe hands.

The other two planes were suffering the same problems and Gemma talked to her fellow disciples as if they were next to her. The pilot was reluctant at first, but finally gave the controls to Gemma who took the captain's seat and the captain stood behind her, ready to take over if he needed to. Instead of heading down, away from the fighter, she pulled the plane up, forcing them to move out of their way to avoid a collision. As the planes got back by the side of the aircraft, Gemma moved the plane to the right and then the left, trying her hardest to push the fighters out of the way. As one pilot tried to cover the plane again with fuel, Gemma turned the plane on its side and the wing tip made contact with the external fuel pod.

Gemma knew exactly what had happened, and using her mind, held both planes together as the fuel was jettisoned up against the fighter's under wing. The fighter pilot was desperately trying to eject the fuel pod and move away from the airliner, but it would not drop from his plane and the planes remained joined together. With the fuel pod devoid of fuel, Gemma allowed the two planes to move apart, but as her plane moved down; its wing tip rubbed against the fighter creating bright blue sparks which ignited the fuel and the fuel pod.

Gemma increased the plane's power and they moved away from the fireball that was now engulfing the fighter behind them. The pilot was staring at the flames engulfing his aircraft and trying to move up and down to put them out but they were getting closer and closer to the engine and his own fuel tanks. Then the outer skin of the wing started to melt and with the heat, ignited the fuel lines inside the wing and five seconds later, the plane blew apart.

Just before the plane exploded, with the pilot still fighting hard to extinguish the flames,

Gemma used her mind to ignite the ejector seat and as the plane exploded, the pilot was sailing through the air away from the fireball. Later he would wonder how he survived as he was sure he had not pulled the ejector handle.

The rest of the fighters moved away from the aircraft as Gemma sent to a message to the other pilots.

"I am Disciple Gemma, I am sorry to have destroyed your other fighter, but I was given no option, if the rest of you do not wish to end up the same way, then leave us now and return to your base. The pilot who I ejected from his fireball, is quite safe and now back at his airfield talking to his Flight Captain trying to explain how he got there and what happened to his expensive fighter."

"This was not our fault, we were only carrying out orders, which we were all against; may Mary and God be with you, we are sorry for trying to intimidate you and thank you for saving our friend Flight Lieutenant Muhammad Icran. This is Captain Sean McCarty signing off, good look Pope One; safe journey," he said over the aircraft's com system which was now working again.

The fighters turned and returned to their base allowing Pope One to continue safely to Rome. With the captain now back in his rightful seat and Gemma keeping a close watch for any other aircraft trying to force them down, the flight continued without any further incident.

With Gemma's instructions, the other disciples did the same and pushed their fighters back to their bases in Britain. Mary could see Mr Templeton would stop at nothing to put an end to Mary and she hoped with what God had planned, it would help her cause.

In her special room, Mary prayed and talked with her Mother, asking for very special help and guidance, like children do at times of danger and trouble. Mary was in trouble, deep trouble, as too was her New Church; if she didn't go carful, then Lucifer would seize His chance and take over the world, then God would never regain control and there would only be one alternative. As Her Mother had said to her before, there were thousands of other planets in the universe to look after, one would not go amiss.

Mary had fought her Mother against the total destruction of the Earth for many years but there was so much sin and corruption here, the planet was infested with lies and treachery. If the New Church folded, woman would perish under Lucifer and man. Man would ensure they would never again be shown up like they had been; man would wipe out the church altogether, every single form of it.

With man's knowledge, he would contact Lucifer and with His help, would continue into the depths of space, where they would carry their war to every planet throughout the Cosmos. There was a chance other planets would become like the Earth Mary was certain, through the men of this planet, Lucifer would find a way to the stars, and destroy what His wife had made so beautiful, would take but a moment to destroy.

Mary could not allow this to happen; she had to think hard and think she did. She spent an entire evening and night, trying to convince her Mother the Earth was worth saving. There were many good people here, and many of the people wished to join her New Church.

There were people who would listen to her she knew, listen to what she had to say. Some people would wish to hear of the lies which they had been fed over the countless years, it was after all, only through Lucifer that the people of Earth grew up the way they had. They had been coerced into becoming the devious people they were.

However, Mary knew if the people of this planet were not saved, they would never be allowed into space. There were too many good life forms out there, to even chance them being contaminated with Earth problems, was a crime in itself.

Mary knew she had to come up with a good idea that her Mother would accept and help her with. If not, except for the few people Mary could save who could live elsewhere in the galaxy in peace, Earth would be a doomed planet. All she could do was hope the people here would build the ships as soon as possible, that way she would have something in which to save those dear to her; and she knew she would save them.

CHAPTER 18

By the end of the night, Mary, although awake the entire time, was not tired; she was refreshed; knowing in her heart she had convinced her Mother, Earth was worth saving. Even in the short time she had been here, it was worth the time and trouble. Starting again in another dimension and closing this dimension down for all eternity, was not the answer.

That was her reason for coming here, for the thousands of years where time had allowed lies to be told, her brother crucified, dead and buried were now over, this was a new beginning; the end of the old times. The times were about to happen in a way people had never dreamed possible. There would be one more night to finalise their plans, while Mr. Templeton was still hiding from the storm, trying to force others against the new God; Mary would agree on a strategy with her Mother that she could work on.

She had once followed her Mother through the depths of space, the boundaries which separate space, dimension and time. Through the non-existent; as of yet, voids which would soon be filled with living worlds, until her energy was almost exhausted, to the far end of eternity itself. There in a zone she had never seen before, was blossoming in young youth, another planet. Stars filled the black night-time heavens, suns, which glared in the night sky, brightening the tiny world, in an immense volume of space with pin pricks of light gracefully lighting the evening sky.

Mary watched, recalling her own thoughts how her Mother took molecules of space, moulded them in her hands, took more and more molecules, adding structure and texture, colour from here and there in the never ending black sky. She spent an hour moulding the materials, until finally, a small globe, with a dent here and there, mountains and valleys, rifts and plains was created. Then quite calmly, as if it were matter of course, she placed it in orbit around the solitary planet and breathed life into the matter She had just created.

Choosing carefully, she picked a twinkling star; she had obviously been nurturing it, because in her eyes, Mary could see the care and attention she had taken over it. The pleasure it gave her with its light; she plucked it from the far side of space and placed it further out from the both spheres and carefully started it to spin.

The three bodies looked splendid, and yet there was something familiar about this scene; she had seen it somewhere before, but Mary had seen countless galaxies, universes and solar systems in different dimensions, space and time. Some even ran parallel, sharing the same space, but separated by time and dimensions as an experiment by Messengerious, another of the great scientists from her realm.

She could not place it at the present time, but she knew she would one day, now she knew where she had seen the almost identical scene before. There would no longer be restless nights, wracking her brain, searching through the vast cosmos of her mind for an answer she knew had to be there; somewhere.

As the thoughts receded from her mind, she visualised the universe and dimension she was in; this planet Earth, with its Sun, Moon and other planets. She had just managed to escape from her mother's sight that day, but had returned on several other occasions; separated by vast amounts of time, to see how it had grown, nurtured, to see if it had expanded like her mother had hoped. What other delights had her mother created? Oh soon, in a thousand years or so she too would take her leave; become a 'Cosmos God' in her own right, create her own planets, galaxies and universes, she would of course learn by her mother's mistakes. She was certain, as a woman, she would not make the same mistake as her Mother. She would definitely not marry, not that there were many bachelors left in their race now.

On her last visit to the prohibited zone; as her Mother called it, she followed her stealthily through a number of mazes which had been created since Mary's last visit. If it were not for her Mother's lead, she would have been lost for sure. Upon arriving, her eyes beheld the most wondrous sight; the entire space was filled with planetoids, asteroids, all in great detail, stars lighting up the dark void of space. Solar winds blew across the endless nights gently pushing small particles of dust which had been blown form planets in their making through space to form meteors; some large some small, but they all now orbited about the planetary system.

Mary was pleased with what she saw, she had herself just completed a course gaining a Masters' degree with honours in 'Universe Structure'. With this course, sub molecular physics and Cosmology 1 & 2 under her belt, she now understood how it all should be and work: how worlds turned within the universe, how they were constructed. Where she would collect the minute atoms from and how to create the first void in which she would fill it with planets and stars. Next she would accomplish the building process, the course was a short one eight hundred years, a very small amount of time for a young Cosmos Goddess to practice the art of planet building, creating the most perfect universe there could ever be.

The universe her Mother had created was magnificent, beyond her wildest dreams; she knew, in a billion years, a billion, trillion, trillion years of learning, she could never create anything as beautiful or majestic as what she had seen before her.

It was art in its finest, beauty beyond beauty; truth with no lies, no deceit and no terror, where only the fairer sex would understand what it was all for. At least the species here would at least have a fair chance, once they had accomplished space travel; passed all the tests, found all the clues, answered them correctly, truthfully and honestly from the heart, they would stand a better chance to see God in the flesh, while still alive, rather than dead and in spirit.

Unfortunately, the people she was trying to help would never solve the clues, yet alone find them if they continued the way they were heading. Mary hoped in their 1960-80's era, they were about to start on their vast conquest, they had landed on the Moon, close to the spaceship that would take them to their next destination, Mars, but alas, no sooner were they on their way when they gave up; foiled by space itself. Foiled; brought down by governments who wished for greater wealth than meeting the creator while they were still alive, yet all this time it was within their grasp.

If they would only have remained on the Moon those extra few weeks, they would have discovered the first key. Of course the Moon was hollow, many of the scientists on Earth knew this; they had just kept it from everyone in fear that it was something else more terrible. The Moon was and still is a vast cavern of planetary wealth; filled with spaceships to take them to their next stop, after of course, they had answered the first questions; they would learn how to fly the ships.

It was easy to her; she had built the ship, set the automatic pilot to its pre-set course, while her brother set the next obstacle on the planet Mars, where a mighty starship, able to exceed the speed of light tenfold and more with ease, was waiting for them inside the face of Mars, all they had to was gain entry to mammoth monument by use of their DNA, so no other species would gain access to the starship on Mars.

Again its auto pilot and computer preset to its next destination, Alpha Centauri, and from there, the supership would traverse the void between galaxies. The next major stop, after one or two minor stops to pick up clues and answer questions, they would arrive at the mighty Andromeda Galaxy. With their clues answered, they would know the location of their next quest. Hidden in a secret location, deep within the guidance system of the ship given to them, were the coordinates for the ship to traverse through the Black Holes, but it had to be the right one not just any Black Hole.

They would enter other dimension, then through to the next and the next and twenty more until finally, at the last dimension, time would be discovered. From there it was one step to see God and receive the most handsome of gifts, the gift of eternal life within life. To be able to play among the stars, visit other nations, form an alliance with other beings. Live life to the full and play among the vast playground called the multiverse. Able to wander aimlessly about, for all eternity; what a future beheld the people of Earth.

Already the Alderians, a lizard type creature, had discovered their third key and were at this moment, travelling to a distant galaxy in search of their quest. All life had the same task, it was obvious as ABC, but the people here had been misled from the very beginning of time, by a selfish man who should have known better. Lucifer, Mary's Father, who she disliked just as much as the governments which ruled the people of the planet Earth.

She returned her thoughts to the *Forbidden Zone*, knowing it was a copy of her Mother's most favourite creation. The Earth system, with beings which resembled them, woman in the same image of the Gods.

Of course if the people who had translated the very first letters in Genesis had added the 's', then the people of this planet would have realised there was more than one God. There were many God's all moving about in their areas of Zone Space.

Then she was caught returning home by her Mother and got the scolding of her life; in an instant the maze was gone and Mary sent home in disgrace, forbidden forever to talk of the *Forbidden Zone*. Where only women existed, no man could enter, yet the women could reproduce feminine beings by splitting their cells. They too had a conquest, but it was confined to the *Forbidden Zone*. Mary had no idea how much inner depth there was to this zone, but there could easily be as many as a million galaxies, each within each other; separated by layer upon layer of dimensions, returning from one space time to another within the same plane.

The maze into the *Forbidden Zone* was made a hundred times more difficult to enter, with alarms at every section, unless you knew the code, the maze changed upon every route; making it impossible to gain entry to this zone. Only her Mother entered the *Forbidden Zone,* but now less frequent, yet always listening to the voices of the people inside; hearing a trillion voices, all asking the same question.

"Why are we here, what is our purpose in life?"

It would not be long before they started their own quest, they were mounting their third expedition to their closest Moon. This time the rocket was big, it was taking 400 females and they had created a proper landing site for their space vehicle. Unlike those people on Earth where they landed here and there, thinking they were following some divine course, but not finding that which they were seeking, because they had not discovered their first clue on Earth, because the bible had been changed. Religion was the last thing on their minds, whereas it should have been the first. Religion is the key, the basis of life itself, but Earth scientists were trying to prove otherwise. She snapped herself out of her thoughts and came back to the early morning.

Through the day, her Mother would back her; she would make a plan which would work. It would make the people, the leaders of this planet among the millions, sit up and listen to what they had achieved and done. In short, they had killed thousands of women and men to save their lives, line their pockets, making sure they were fine, while millions starved and fought to survive; how selfish could the world leaders be?

Mary came out of her thoughts and plans again when she met the others at breakfast. Everyone was refreshed, but none had thought so hard as Mary to save the world from her Mother's wrath, luckily for them, they knew nothing of what her Mother was planning to do.

They discussed their plans of what would happen that day, but no one understood what would happen to their leaders; how they displeased Mary and God so much. After breakfast, they departed on a tour of the local churches of all denominations; their aim was to get these people to join the New Church, to put themselves on the right road to the salvation of God.

Work was progressing on the structure of the Space Arks and the buildings which other people would work in. More souls from Heaven were interested in this planet's future than first thought and help was coming from everywhere.

The day's visits went swiftly, with many church leaders agreeing as soon as they entered the church doors to join the New Church. Mary was welcomed by church ministers wherever she went; it appeared only the politicians had not seen the light, well some of them, even when it shone so brightly before their eyes. They were blind and lost in a world that could easily be destroyed, not now by themselves, but a far higher power of government.

Once agreed and the formalities taken care of, the next task was to inform the people, which was left to the ministers who had also become disciples of Mary, if the disciples couldn't accomplish this, they would not be worthy of their positions. Many of the disciples called from house to house getting everyone to come to their local church where they explained to everyone in one go what Mary stood for and what God wanted of them.

By the evening, over three hundred small churches were converted, all the ministers wondering when they would get their new church. The congregations gave the ministers their full support having been talked to by the disciples and this was to the great annoyance of a few government ministers. Even Lucifer was slowly getting to His feet, aware of what was happening on Earth. He was gradually losing His hold over the people he controlled with money and material objects. He had ruled the Earth for so long; He had taken it to be His, despite it belonging to His wife.

In South Africa, the hurricane was moving away, and word was getting through to those in the shelters that in a few hours it would be safe to return to the surface. Already, the emergency services were on the ground, looking around at what had happened to the surface and where the storm was now heading, further north and hitting landfall again in another highly populated city but those people were already underground, ready to sit out the storm. Nobody had waited to hear what damage had been done in Johannesburg.

As the people climbed back to the surface from their underground havens, they were amazed at what they saw; it took the people time to come to terms with the change to their city. The one shelter that had its entrance covered in heavy boards was cleared and the people allowed to return to the surface. When the first news travelled down to the underground station that the people could now return to the surface, Mr. Templeton reminded everyone what he thought of Mary and reminded them that it was God who sent the hurricane to their city and destroyed their homes and lives, not the governments, it was up to the people to join him in a show of strength and bring Mary down and send her back to Heaven.

As the first people walked safely onto the surface again, the people left inside could hear the cries of everyone above. Mr. Templeton smiled and again used the time he had as he walked up the thousand steps to the surface, to get people annoyed at God and Mary. When he finally got to the surface, smiling and joking with his aides, he looked around at the destruction of the city, not believing his eyes. It was just as well everyone was underground, the storm had been very destructive, it had destroyed everything in the city, and God had come right in behind it and rebuilt every house, shop, Mall, factory, government building and most all, the churches. The roads had been rebuilt, wider, safer, the drains were repaired, water, electric and gas was in every house, it was brand new and all free.

Every house was given a large television in every room; all the latest computers, again in every room, free internet, free TV stations, TV companies had thousands of films from all over the world and could play them to everyone. As soon as the people arrived on the surface, they knew where their home had once stood, their new home would be, they were all given jobs, there was money in their bank accounts and nobody was sick or injured. There were no unsightly wires going from post to post carrying electricity or telephone lines, it was all underground and there were underground walkways throughout the city for people to service the cables and ducts.

Two camera crews drove as fast as they could up the coast and followed the hurricane, seeing it and passing the film to every station they could. They showed the storm flatten the next city, then God's Hands rebuilding it, making it all brand new.

Mr. Templeton looked around and couldn't believe what he was seeing, he was not expecting this and neither was anyone else. The people were down on their knees, crying aloud and praying to God, thanking her for what she had done to make their city look so beautiful. Four disciples appeared in the middle of the road and talked to everyone. As Mr. Templeton tried to shout them down, he was picked up by a car and taken to the nearest airport where his plane was refuelled and waiting for him and his people to continue with his journey. Before he left, he met with one of the church leaders.

"I don't know what God has tried to do here, but She has obviously attempted to show me up. I urge you not to believe in Her, this was not a miracle from God, it's all an illusion, created by God to make me look small. I bet when I leave all you will find is devastation," he said in an angry tone.

"Mr. Templeton, I have spoken to the disciples and they have assured me and everyone in the city this has all been done by God to help us. I for one believe in the new God and I for one will not speak against Her and neither will my people. I suggest you change your ways and believe in the New God before She takes your life and changes it forever. I am telling you, on behalf of the people of Johannesburg, do not return to our country again unless you change your ways, we do not wish to have sinners here; especially when they speak against God, who has obviously done nothing but good for us. Goodbye Mr. Templeton, do not return here again until you have a better heart and have repented your sins. If your plane runs out of fuel or breaks down, you will no longer be allowed to land anywhere in South Africa or North Africa for that matter."

"I am sure you are wrong about God, I have personally met Mary and I can assure you, she is not what she appears to be. One of my minsters who admittedly stole money from people and did things wrong in government to make money for himself, was turned into a cripple by Mary before everyone including the Pope and the King and Queen of England. She gave him emphysema and turned him into a cripple, not even his wife and mistress helped him. The King gave him the sack and I had no option at the time to fire him as well, but I will ensure he's okay. Mary can be very devious; don't trust her, you might find yourself in wheelchair as well."

"Actually, he is only going to be in a wheelchair for one week, although he doesn't know it yet, he has also been stripped of his money and his family have been given more than enough to take them through their lives. He is currently being spoken to by people who only wish to help him and are used to working with disabled people and he is thinking of doing something to help others when he gets used to his wheelchair. He has already been praying to God for forgiveness and nobody has pushed him into it. I believe King William has offered him a job as soon as he is used to his new condition. I thought he would have told you," Father Graham replied.

"I noticed he has tried to call me twenty four times, but I was talking to other ministers and could not take his call. Anyway, farewell, no hard feelings I hope," he said smiling and held out his hand to shake Father Graham's hand.

"I'm sorry, I cannot shake the hand of a nonbeliever, please, speak to your friend, he wants to tell you something very important. Goodbye Mr. Templeton, do not return here unless you're with God."

Father Graham turned without shaking his hand and walked away from the airport terminal getting into his waiting car and returned to his new church to pray with his new congregation to God, the female God he now truly believed in.

Boarding his plane, the door closed and he hoped there would no further problems on his trip to Australia, where he was determined to put the people to right. When he contacted the Royal Air Force officers who said they would do his bidding, he found they had not managed to destroy the other aircraft and the officers would no longer speak to him and were now carrying out other duties so they would not be able to assist him further with black operations.

Mark was still very ill after God touched him with Her hand; He was cleansed, which Lucifer hated the most. All Mark would talk about was seeing his sister and brother, he wanted desperately to meet and talk with them, after being locked away from his Mother since birth.

The group of disciples and Mary talked in high spirits, making two further calls during the early evening to more churches. Later, thirty five vicars arrived with their church helpers for a late supper to discuss the New Church. They too went away pleased with a change of heart; they even agreed to take more responsibility for the church, look after their children, so the women could learn more about the church, especially the fifteen Mary chose to be disciples; to teach other women about God and the bible.

Her group of disciples were learning quickly how Mary's tactics worked. Unfortunately on television, the government was still not a hundred per cent behind God. Mr. Templeton was still the Prime Minister for the moment, but upon his return, if not earlier, he would be told his position was no longer required.

At one church they had to help some people because of what they had been told by Mr. Templeton's people. "I am not asking you for a pledge of any form, you do not have to pledge your wages or your wages of sin and belief. You do not have to pledge any property either, God will take care of the churches She builds and the running of them," Mary said honestly.

"But we have been told by the Prime Minister's people our church will no longer be assisted through the Church of England, we will get no financial help or wages. The letter continues to tell us we are expected to pledge our private property and our own money to your movement to help support your vast organisation. If you have no income, how else will it operate?" Cannon Mason asked Mary.

They were all shocked upon reading the letter from the Prime Minister's office. Now they understood clearly what they were up against.

"This letter is a farce, it is written in anger, trying to stop you from changing your beliefs. Mr. Templeton wishes to remain in power of the people and the Church and put down the Monarchy and eventually, he will take control of churches that do not follow us and take their wealth and gold," Mary explained as calmly as she could.

"God will take care of all the minsters and Her churches, do not forget, everyone has started new, you have all been given a second chance. There will be no more chances from God, this time, you either believe in Her or not. There will be no Third Coming, we have no more brothers or sisters waiting in Heaven to come here in another two thousand years.

"If you want to start a fund for a children's sanctuary that is fine by me, but I am not asking for your money or property," Mary told an anxious group of people at Silbury Hill.

"We did not mean to offend you or God, our intensions are honest, please believe us, it is because we have been given this second chance; a second lease on life so to speak, that we feel we should give something back in return," Mrs. Leslie Bullock, a church Governor said apologetically.

"No offence taken Leslie, like I said, if you wish to do something, set up a fund for anything then it is up to you, this is one thing my Mother has not covered but by the time My work here is completed, there should be no starving people, no homeless or need for people to help others, except in time of a natural disaster, which hopefully will happen less and less as God is accepted," Mary paused, having had a wonderful idea; here was a lady she could work with.

Without thinking further, she looked deep into her mind, what she saw made her heart weep. The poor woman had lost two of her own children in a hit and run accident by a drunken driver the man, only 19 years old was until a week ago serving a five year prison sentence. She could see for the victim's relief, something had to be done to justify their continuing to live. This woman's heart was pure, as too her husband's who both missed their children dearly. Mary instantly formed a new section of the church and Leslie Bullock was to have the shock of her life in the next few minutes. Mary would give her something she was not expecting and Mary knew she would accept the position with all her heart.

"Leslie and your dear husband Colin," Mary said. Both were shocked at being called by their first names. "I would like you both to head a department of our church for children; it will be up to you to find homes for all the children and babies that are without their parents in this country. When your job is done here, you will start abroad," Mary explained shocking everyone in the room.

"But Mary," Leslie said at last. "We are both middle aged, surely someone much younger would do a better job, and we have no experience in this and neither of us speak any other languages."

"You do now," she replied placing her hands on their heads'.

A moment later they spoke every language of the world and more they would use in the years to come, proving there was an Earth still here.

"You will work from your church when it is shortly rebuilt, as for the moment, you will take a short break here at my mansion while you are instructed in your duties and everything is set up for you," Mary explained.

Everyone applauded them and at the end of the meeting they were taken by one of the guides and shown to their new rooms. Everyone else was taken by coach back to their church where they departed for their own homes.

Back at the Mansion of Disciples, when the last of the disciples retired for the evening, Mary went to her own room. Her mind was bursting with ideas, thoughts of what should and could be done to make the government see where they had gone wrong. How they could help and move both woman and man into a new life which was so rewarding and fulfilling.

God had been thinking too, despite Her endless day's tasks, seeing and listening to endless voices of cries for help. Pleas, pledges of the soul in exchange for a daughter or son's life, She could help some but not all; many had to find their own way through life, sort out their own problems, but the government was a priority and so She put a great deal of thought into their problem and how She could help her daughter.

There was also an underlying problem which was far greater than the government, Lucifer Himself. Until woman had ventured into space, they were prone to His attacks; that was why, She had passed by the Moon and Mars quests in one go. She had given the designers so much to do and understand.; it was not a question of just drawing the ships, anyone with an ounce of imagination could do that, but they had understand the maths and principles of flying the ship through space and these were very large ships.

Mary and Her Mother talked about their ideas far into the night, slipping through a time zone, they talked even longer until Mary returned to her own time to a point just before dawn. They were now agreed upon the way to move forward and decided the following day would be a good day to start. It would give the government time to relent, to come to terms with the changes which would take place and to accept the New Church for what is was and stood for, the salvation of woman and man and most of all, Earth itself. It would give Mary time to create the perfect zone for the purpose in hand.

The following morning they arrived at their first destination, Sideral Church in Oxford. They were met by church officials who greeted them with shaky hands, they looked ill at ease, worried and weakened by interrupted sleep and Mary knew something was wrong.

"Mary, supposedly the daughter of God, you are wanted for questioning by the Government Supreme Command," a man said from behind the group of ministers in a deep voice, stepping through the group of men. He held Mary's arm tightly, slapping a set of solid chrome handcuffs over her wrists and smiled grimly then looked to his colleagues who were arresting the disciples at the same time.

"We have a van to take you to government headquarters where you will answer questions which are put to you." The man said sternly. He turned and nodded to the man by his side and tried to pull Mary forward.

She stood her ground and the handcuffs slipped off into his tightened fist. This was followed by a clatter of metal on the cold stone floor in the outer room of the church. The officers looked at the handcuffs in dismay as the tightly fitted handcuffs dropped to the floor from the disciples. They all smiled, looking in a forgiving nature at the officers as one man pulled his gun.

"Move or I'll blow your bloody head off," he snarled at First Lady Rocc.

She looked frightened at first, brought her hands to her face, as if about to cry then gripped the gun tightly and started to eat the chocolate gun.

"Would you like some chocolate, I must say it's the finest I've ever eaten?" she asked the man who now looked shocked and petrified as she held out half of the chocolate gun. The man snarled and knocked the chocolate from her hands which fell to the floor and broke into several pieces. Instantly a number of rats and four mice came scurrying out from beneath a cupboard and consumed the chocolate and looked with menacing teeth at the men standing before them. There were still two small pieces of chocolate on the floor which the mice had left as they all had their fill. First Lady Rocc bent down and picked them up, raised her arms above her head and blessed them.

"You," she commanded to the police woman in uniform at the back who had now shown herself. "Take this to the local infant's school, you will find all the children in the school hall praying. You are to stand at the door of the hall and give a piece of chocolate to each child the size of your thumb. You will find the two pieces will quite easily go between the two hundred twenty three children in the hall and there will be some left for the staff."

She looked down at the small pieces of chocolate and wondered how this could happen. She turned her head to look at Mary who nodded and smiled back to her. The woman officer left without a further word and dismissed the objections of the male officers who shouted at her but could not move their feet or arms to stop her leaving.

"We have a meeting with these good church ministers, after that meeting, we will accompany you, you may send the van away, it will not be needed, we will take you all with us to 10 Downing Street, but I will not submit myself or my disciples to travel in that decrepit van like terrorists," Mary informed them in a very stern voice.

She walked past the officers followed by her disciples and ministers, the men stood perfectly still, unable to move as a cold wind swept into the church, whirling around them, making them feel colder and colder. The officers could turn their heads and saw the van leave despite their cries for help.

After two hours of talks inside the warm church, the church ministers agreed whole heartedly to accept Mary's offer and join her New Church. Upon the spoken words "Thank you" by Cannon MacAuley representing the church ministers and governors, the ground trembled beneath their feet.

"What is happening?" they asked thinking it was earthquake.

Mary smiled as too did the disciples. "It is God repaying your thanks. Let us walk into the main church," Mary said calmly.

Leaving the vestry, with its old furniture, they entered the main church, where hours earlier lay broken pews, uneven floors and a pipe organ which refused to work. A piano stood in the corner, not the best instrument for leading the few people who attended the church in Morning Prayer, but it was better than nothing.

As the ground ceased to tremble, the floor was even, covered in a red carpet throughout. The pews were all new, not the old fashioned wood, but in teak, they were also padded with a cushion the entire length in red velvet trimmed with gold thread. The piano was gone, now completely refurbished it was returned to its owner. In its place, the organ was humming away apparently playing by itself until the organist popped her head from behind the new curtains which hid her from view.

"It's fantastic Cannon MacAuley; just listen to the sound of the organ, it's like listening to music from Heaven itself. The church as well, just look at it, isn't it wonderful, people will flock here just to see what God has done and they will continue to come, despite what the government is saying. They will come, I'm sure of it," Gladys said and continued playing the organ like she had been playing it for a hundred years. Music just flowed from her mind, not a note out of place.

Above, in the gallery, it was just the same, completely refurbished throughout, as like the inner walls, now brightly painted in white and cream, with lights twinkling like candles from their holders, now disconnected from their mains supply with all the paperwork completed and filed away. The heating was up to date and working full blast to heat the church and bring in the people from the cold outside, again the gas boilers disconnected from the mains and the fuel supplied by God.

"The church will remain at a constant temperature from now on; it will no longer be a cold place in which to worship and your church will be filled to the rafters and more people will come, have faith Cannon MacAuley," Mary said.

"But the fuel costs and electric bills?" muttered Cannon MacAuley.

Mary chuckled. "You have been disconnected from the mains services, gas, water and electricity. The waste for the toilets will continue to be taken off by the water companies, but I have arranged for all these bills to paid for the next hundred years. All your heating and lighting is free so make the most of it and welcome your parishioners here day and night."

Cannon MacAuley's eyes lit up with excitement.

"The same goes for the vicarage and all the buildings connected with the church," Mary added, "It's all in here," she continued and handed him a small pamphlet which appeared from thin air. Cannon MacAuley took it with eager hands and flipped through it; suddenly realising he could read very fast and recall everything he had read.

Passing the outer hall of the church, the wind was slowly dying, but the officers still could not move. Outside, the bells which had been taken away many years ago were replaced, pealing gaily, brightening the morning air. The church was completely rebuilt around them, a different shape, size and roof; it was of the typical New Church format.

"It's magnificent," everyone said together.

"It's criminal what you did," a voice shouted from behind them. "I'll add it to your list of crimes," the officer shouted.

"Saying what, you were held hostage by the wind?" Mary asked calmly. "Come now, it will be laughed out of court as well you know."

"When the van gets back we're leaving and you are coming with us even if I have to beat you all up and break your legs," he shouted in an angry voice.

"What is the meaning of this?" the new Prime Minster, Mr. Paul Ashley demanded to know as the police officers, shouting at Mary and threatening her and her disciples appeared inside 10 Downing Street.

The four police officers were aghast at what had happened, one moment they were in Oxford inside a church, the next they were in the capital. There was no sense of movement, time lapse or sped, it was amazing, fantastic to say the least; it took them a full minute to recover from their experience.

"I repeat, what is happening here and why are you shouting and threatening Mary, the daughter of God you imbeciles?" Mr. Paul Ashley demanded to know as four security guards entered the room.

"Oh it's you Mr. Ashley, the Prime Minister told us to arrest these people and bring them to headquarters where they would be interrogated under drugs to get the truth out of them as to what they were intending to do to us. We have been told to exterminate them and use everything we have at our disposal. You have no authority over us, we only work for the Prime Minister, now, we'll take these terrorists to where they deserve to go," Commander Ian Blake shouted.

"I am now the Prime Minster, a vote of no confidence was taken for Mr. Templeton and he has been sacked from his post under public demand and has been recalled back to England. Unfortunately, he was in Australia when I told him the news, a little like the way we did it to Margaret Thatcher when she was away in France. He was deported immediately with all his staff and put on an aircraft to get him out of the country as swiftly as possible as the people in Australia were shouting and getting ready to lynch him.

"As the plane was made ready, there was a mix up with the fuel and the plane was only half filled. Someone didn't change the weight from pounds to kilograms, as it was, the plane was only half filled with fuel and although the captain tried very hard, no country would allow him to land and he was forced to continue his journey across the Atlantic Ocean."

"What about refuelling the plane in flight?" Commander Ian Blake shouted in reply.

"The plane did not have the capability to refuel in mid air, you need a special adapter for that and you have to be trained to refuel your aircraft in flight. The pilot and co-pilot were not classed for this procedure anyway. As a result, the plane ditched in the middle of the Atlantic Ocean, a number of ships were sent out and the pilot did try to find a ship to ditch close to but the closest ship was two hours away. By the time they got there, the plane had gone down and only the pilots and cabin crew were saved. Apparently, there were a large number of sharks in the area, so Mr. Templeton met his friends so to speak," he explained grinning.

"As for you three officers, you are all sacked from your posts, you will hand your weapons into your superior officers and clear out your desks within the next two hours. My officers will take you to your offices; I suggest before you leave you ask Mary for her forgiveness, I can assure you all, you do not wish to meet the same fate as your previous Prime Minister. It is surprising the number of sharks there are around at the moment."

Just as the men were about to be led away, Mr. Spencer, MP for Internal State Affairs thought to himself.

"Before you dismiss these officers, who were obviously working under Mr. Templeton's direct orders, were only carrying out their duties. The other thing we need to take into consideration is the fact that the government has done so much behind the people's backs, taken money for expenses when we shouldn't have and of course then as General Buck here will admit I am sure, there is the problem with wars."

"What about wars?" Mr. Paul Ashley asked.

"To be honest," General Buck intervened. "We have been with a few other counties, making wars happen to keep the number of lower class citizens around the world down and to test the new weapons we build. How else are we to test them in the field in peacetime, so we need to create wars and terrorists to test them? Surely you don't think the terrorists actually blew up the Twin Towers back in 9/11/2001, they would never have collapsed like that if engineers had not removed the main supports three weeks before, it was lucky the buildings remained standing until the planes hit them.

"It wasn't just the planes either; we had to test a new laser, a very powerful laser to see if it really worked, we were so pleased with the outcome. Then it was cat and mouse with the terrorists who we found, let go, found again, killed a few of their people, they killed a few of ours and we all played at testing our latest weapons. It is how we improve them, like playing paint ball games in the country, only we use real weapons and we give both sides the weapons and see who comes out on top. The leaders of the country only get to know the truth if they need to know. Take the bus explosion in London on 7th July 2005 and the underground train explosion, we had to blame someone and they were only black people. We planted the bombs ourselves on the buses and the trains, we needed people to demand we take action and take action we did."

"I don't believe what I'm hearing, how could you kill our own people like that, kill Americans in their tens of thousands in the Twin Towers?"

"They didn't all die, we told a lot of people they were not to go to work that day, so we saved those people that were needed to be saved, and it was mainly coloured and Mexicans that were killed, again the lower class citizens."

"We don't want you to build your spaceships here in this country," Mr. Spencer said plucking up the courage to carry out his former Prime Minister's wishes. "It won't do us any good."

"Well, we'll move our spaceports to another country if you don't want the jobs filled and your people back in full time work. These spaceships will give over three million people honest work and it will be well paid work as well. I wouldn't like to upset your plans for world destruction. However, if you really wish to destroy the world, my Mother can do this for you right now; it will save you a fortune in atomic weapons and men. If you give me the information needed, I'll see it is done," Mary said with a smile.

As Mary spoke, she sent a calming message to her disciples.

"What information are you talking about?" Mr. Spencer asked, he sounded very confused at the situation that was now being dumped in lap as Mr. Templeton was most likely dead. But he knew Mr. Templeton well, if he did get back alive, he would want to know why his plans had not been carried out, Prime Minister or not; he hated Mary for showing him up before the world.

"What country would like wiped from the face of the Earth first?" Mary calmly asked, just as if it were an everyday question carrying no real meaning.

"I . . . I . . .I . . . I . . .I," Mr. Spencer stuttered.

"Come on man, the answer is bloody obvious, the Russians are our greatest threat, have them blown away first," General Buck bellowed, standing up from his seat and banging his fist onto the highly polished, oak table.

Mary looked at him and smiled. "Do you mind telling me, word for word what you would like done, so I know what it is you actually mean," Mary continued.

"Bloody hell, are you thick? Women ehh, you have to put words into single syllables, listen to me lass; I . . want . . you . . to . . blow . . Russia . . off . . the . . face . . of . . the . . fucking . . Earth . . . is that plain enough for you?" he said slowly, word for word.

"Exactly, I just wanted to be sure before I asked my Mother to do this for you."

Everyone looked at each other as Mary closed her eyes and thought for a few seconds. "Please observe the wall screen behind me."

A huge wall screen suddenly appeared, suspended on the wall and showed a picture of the Earth as seen from space. The camera zoomed down at horrendous speed making everyone feel a little queasy as it showed Russia from above; then it was like they were on the ground, actually in Russia as they could see the Kremlin and the people were dressed in Russian clothes as it was their winter.

Everyone was looking up to the sky, then people started to run and scream in fear and terror as huge fireballs hurled down on their city. The explosions, as the fireballs hit the ground made the cabinet room shake. Everyone was shaken as they were not expecting this. They could hear the loud screams and even smell the fear in people as they ran for cover from the descending fireballs that landed and exploded all around them. Buildings were crumbling to the ground, a crowd of people burst into flames before their eyes; the smell of burning flesh filtered into the cabinet room and into the noses of everyone there. It was just like they were in Russia at the same time. There were screams and shouts, calls for help all around them; the sound of people running and buildings plummeting to the ground, people were coughing, choking through the acrid smoke and debris. As three people looked up to the sky, a huge sheet of glass came down and decapitated all of them at the same time, blood going in all directions and their heads rolling down the road, their eyes still open seeing the buildings about to come down upon their severed bodies.

It was disgusting to say the least, then, those in the cabinet room tried cover their noses to hide the obnoxious smells but were unable to move their arms; they had no option but breath in the fumes of death. They also had no option but listen to the cries of pain and death as they happened to the people in Russia that very moment. A printer burst into action, the secretary was allowed to walk to the printer and take the paper from the machine and hand it to Mr. Spencer, who was allowed to move his hands, but not cover his face. The red phone on the Prime Ministers' desk rang and he picked it up.

"What have you done to us Prime Minster Templeton?"

"I'm sorry President Dmitriev; Mr. Templeton has just been removed from power and I, Mr. Paul Ashley am now in control of the country. Tell me, what is happening there," he asked.

In the background, he could hear 'Rule Britannia' being played very loud in the sky. Flames licked the country, burning the land, people and animals alike with no mercy. Mr. Ashley couldn't reply he didn't know what to say and neither did General Buck, who was suddenly taken from the room and appeared on the screen. He was there in Russia, in the thick of it, witnessing the total destruction of the country he had asked to see blown from the face of the Earth. The cries of pain and anguish, the blood and tears, the searing heat scorched his hair, despite the protective suit Mary had put him in. The smell of death was everywhere, including the cabinet room, it was sickening.

Then General Buck was back, trembling in his seat, his face blackened with soot, in deep shock, frightened to death. A view from space showed the final scenes as an enormous section of Earth, was torn from its resting place. Russia was rolled up like plasticine and compressed into a ball and thrown, complete with bodies, some still half alive, far; further still to the distant stars.

The smell and taste of death, burning flesh lingered on, the red phone was now dead, the printer ceased its tapping operation and the room was completely silent.

"What has just happened?" Mr. Ashley asked Mary full of trepidation.

"What is all that white smoke where Russia was?" General Buck demanded to know pointing to the large wall screen.

"Oh that, it's nothing to worry about; at the moment: it's just the sea pouring into the abyss where Russia used to be. You have to realise that Russia was a huge continent; the entire continent was removed down to the magma core and it's an awful big space to fill. By my estimates, and I may be out by a few million litres, but it looks like the entire worlds' oceans will pour into the abyss of former Russia. I'm afraid this is the side effect of taking one person's word to the very extreme, which is why I asked you General Buck, to tell me word for word what you wanted to be done.

"Had you said you wanted Russia flattened, then it would have been flattened, but you wanted it blown off the Earth and into space, which is exactly what God did for you. In less than three minutes, she killed," she paused, "One point eight nine seven billion people, over ten million animals and trillions upon trillions of bugs. Just to confirm they are all dead and gone, and the country is as you wished, now flying through space at two hundred times the speed of light. It is on its way to the end of time, and currently irretrievable," Mary explained.

"Forgive my ignorance, but I don't quite follow," Mr. Ashley said trembling in his seat.

"In simple terms Mr. Ashley, I did as was commanded by General Buck, word for word. God blew Russia from the face of the Earth. You see, the Earth is about to be destroyed; he obviously did not think his request through before asking it. But there again, that is what he wanted for his boss Mr. Templeton, who is, I assure you, now dead and eaten by three sharks.

It is what you asked for is it not General Buck?"

"Why yes, but I. .I . . . I . . ." He could not conclude his sentence, he could see the destruction of the planet he was part of before him. The Earth was getting ready to blow up. Everyone was silent; they never thought it would happen like this.

"What will happen next?" Mr. Ashley asked.

"Ah," Mary paused and looked at her watch and then at the screen. "It is about time for my disciples and myself to leave; you'll find out soon enough what will happen next. However, to enlighten you as your desire to know is of the greatest importance to you. Sea water is currently spilling into the abyss that was formerly Russia; the Arctic Ocean has been turned to steam as it fell into the molten rock mantle beneath Russia. The North Sea is still pouring into the abyss and being turned to super steam and the super steam is devouring more and more water as I speak. You in short, will all die! Everyone on Earth will die, even inside bunkers, no matter how deep, or what they are constructed from. The superheated steam will find you and melt you to death."

"The nuclear bunkers, we'll go there," Pauline, the secretary suggested in a terrified voice.

"The steam will melt the concrete and lead shielding, you will suffer even more, it will be a worse death than if you walked into the steam right now," Mary stated.

The frightened men, once powerful, found themselves crying for help from Mary. She shook her head in dismay.

"It would mean creating this planet again, I'm afraid not, this is what you wished for, is it not General Buck?"

"Yes. . . but. . . no . . . Not the women and children, not to die like that, not for us to hear their screams right here in this room. It's . . . it's disgusting; that happens in the front line, it's for those men out there to see, not us, we're politicians, we don't get involved in that end of the war," General Buck stated annoyed and frightened.

At that moment, the screen showed a scene from Britain, it was Birmingham, they saw the Bull Ring go up in smoke, then people started to catch fire, a huge ball of white hot steam covered the entire area, dissolving the concrete and soil. People were screaming their lungs out as they died from steam burns, the stench again infiltrated the room and the screams echoed around them.

"Disgusting you say, wait until you feel your own skin burn, smell your own flesh as it catches fire, hear it crackle with the heat and melt, drip from your bodies before your own eyes. Your skin and muscle will melt down over your bones and God will ensure, to allow the billions of souls into Heaven that die here today; you will pay with their pain and suffering. For all of you, including your two women secretaries, you will live and experience the time your body is dying," Mary stated.

The men felt sick to their stomachs and sat down in disgust, they looked horrified at General Buck for demanding Russia to be blown of the face of the Earth. It was all a mistake, a huge enormous mistake, they knew they had to get that message through to Mary and hopefully she would get it through to God.

"Please, if you can reverse time in some way, you are after all the daughter of God. Please, I'm down on my knees, I beg you, for all the people in the world, for the lives of the people in this room and all over the planet. Please, somehow reverse what has happened. Please!" Mr. Ashley prayed and begged on his knees, clutching his hands together until they were white. The Prime Minister knelt low, prayed like he had never prayed before, for forgiveness and salvation, not just for himself but for the entire world he and his party were about to destroy.

The view on the wall screen was changed to show the Earth as it had always been, Mr. Ashley slowly got to his feet, picked up the red phone which was answered within two rings and spoke quickly.

"Are you there Yuri Dmitriev?" he asked frantically.

"Yes, I'm here, what are you trying to tell me, are we going to war?"

"No. . . No, nothing like that at all. God be praised that you're alive and safe. Thank you for being there, I am so pleased to hear your voice. You must come to our country for a state visit very soon, please bring your wife and your advisors, and we will join our two countries together forever more and be friends, very close friends and no longer be at conflict with each other. All of your people are welcome to come to our country whenever they like. Please, tell me you forgive our stupid ways of the past and help us to move forward in peace," he asked with a strained voice.

"I agree, we must move forward and with Mr. Templeton gone from office, I can see nothing stopping us joining forces and moving forward in peace and harmony and helping each other. Your people too are most welcome in our wonderful country. I will get my secretary to make the arrangements for us to come together very soon. Thank you Prime Minister Mr. Paul Ashley."

"Please, my name to you is Paul, and I hope we can be friends for the rest of our lives."

"I see no reason why not, I look forward to seeing you very soon," Yuri Dmitriev said and replaced his receiver.

"Thank goodness they are alive and well, I think I have made the first steps to join our two countries together in peace; I hope so anyway. It's about time we thought again about war and what it means for us. I suggest General Buck, you think before you speak in future." He turned to face Mary. I am very sorry about what has been said and done, please, continue with all our blessings with your New Church, you may convert those who will listen to you, I will not pass any laws to stop people from changing to your religion and I will of course give you all the assistance you need. I will also ensure whatever God builds or rebuilds or alters, the councils will not interfere with anything and help you with whatever you need.

"I am sure if God has built the buildings, they will be much safer and last much longer than anything we can build ourselves, She knows exactly what She is doing, after all, nobody put buildings up overnight without some form of miracle and She did make the Earth and stars, they have remained in place for countless millions of years, I am sure her buildings on Earth will do the same. Thank you for reversing the devastation which has just taken place and bringing our two countries together. I have learned a great deal from this experience, we must all be very careful what we wish for in future, and choose our words very carefully; least we get more than we asked for," Mr. Ashley sighed.

"It was you who joined your two countries together not I, that was all your work and your words, which just goes to show how a few words can end years of hostility. It is about time you saw firsthand what happens in war, we are leaving now, but before we depart, I take it we can still have our spaceports in the UK?" Mary asked.

"Of course, have and build whatever you wish, all I ask, is that you will allow me at some time to help you and visit the shipyards. I would very much like to see one of the ships, inside as well; as you probably know, I am very interested in space research and spaceships in general," Mr. Ashley replied.

"I do, and I will see to it you get an invitation to go over the ship during its construction, not just before it goes into space. Thank you for your time; have those three men taken away and send them to the shipyard tomorrow in Russia with their families if they wish to accompany them. I'm sure once in Russia, they will see a different sort of people who are now at peace and only wish to help and pray to God," Mary said.

"Russia, but how will we communicate, we don't speak any Russian at all?" Commander Ian Blake objected.

Mary placed her hands over the heads of the three agents and they felt their bodies start to tingle.

"You now have the ability to speak any language on Earth and any species you will meet in space. You also will know exactly what to do in helping to build the Arks and you will be on the third Ark to go into space with your families, if they wish to go with you. Now, clear out your desks and arrive at Heathrow Airport tomorrow at 07.00 for your plane to Russia," Mary explained.

The three men felt a lot better and more relaxed as they left 10Downing Street.

"Thank you once again Mary for all your help with putting this government to right. My door is always open to you," Mr. Ashley said honestly.

Mary joined her disciples and a moment later they all disappeared, going to their next destination, where the church leaders and their coach were waiting for them.

CHAPTER 19

Over the next five days things calmed down with politicians all over the world, but there were still a number of leaders, especially in Islamic countries where the men could not come to terms with allowing women into the church to lead them in prayers. As there was so much hope for reform in the UK, Europe and Russia, Mary concentrated on these churches first.

After one week, Mary sent her top disciples out on their own, now in a slightly safer world; First Lady King with a hundred disciples went to Russia to prepare the people there. They took with them a coach and ten cars, all with glass tops so the people could see them as they moved around. First Father Langley was given the job of converting every church in the UK to the new church and with fifty disciples, he started his job.

Second Lady King, First Mother Meadow, Second Father Huntley with another four hundred disciples were sent all over Europe to bring more and more people into the new church.

Father Mathews, with his knowledge of Islam, was sent there with only fifty disciples to try and bring about a change of faith and to explain to the church leaders there what Mary and God expected of them. He would take many disciples as it might overwhelm their religion and make it appear they were attacking it, which would not have worked.

Mary was getting fifty disciples to take with her to Ireland, joining the north and south together for the first time in many years. They had one week to bring about the change and First Father Langley suggested that if they used a large helicopter, they could assist people in difficulty and also have the new church sign all over it so people would become accustomed to seeing the helicopter flying low over the country and ask more questions about the new faith. As First Lady Rocc was arriving in Russia, Mary joined her on the plane and spoke with her for a few moments before taking her own seat and deplaning with the group.

When Mary and First Lady Rocc arrived in Russia, they flew in on a normal plane and were met by hundreds of press and President Yuri Dmitriev. It was a joyous occasion as President Dmitriev told everyone via television that it was Mary and First Lady Rocc who had saved their country from destruction and brought about peace between Russia and their Western Allies.

"My fellow citizens," President Dmitriev said before a large number of television cameras and press officials, including Mike who had joined First Lady Rocc on the trip to report her trip for his own TV station.

"I would like to introduce to you my very good friend Mary, the daughter of God and First Lady Rocc, the leader below Mary, of the new church of God; Jean, to her friends and I hope I am allowed to call her by her first name," he said looking at her.

She smiled and nodded in the affirmative.

"Mary and Jean are here to convert you all to Mary's new church. Now, I know we have believed in God, which I am told is also good so that we have a faith, even though it was the wrong God we were following. I want everyone in this country and our allied countries to turn away from the God we once worshiped and follow the new God, the true God who although a woman, when you look at what She has created, it obviously had to be a woman, a man could not have thought up all this beauty. Man can build a house, but a woman makes it into a home. It is the woman who adds all the detail and makes it habitable and a nice place in which to live.

"Saying this, I give my full permission for Mary, First Lady Rocc and her disciples to travel wherever they like and talk to whoever they like. You will treat everyone in the group with the greatest respect and listen to what Jean and her disciples have to say. I will not force you or tell you to join their church, it is up to you. From this day forward, the KGB is being disbanded, the special police forces are also being disbanded, they now have other jobs to do. The army, air force and navy will remain, but these people will have a different job. The air force will use the helicopters and aircraft to ferry people around our vast country and where accidents have happened, the helicopters will be adapted to carry patients to our hospitals, which I hope after speaking with Jean, she might be able to help us with.

"The navy will be used to help transport people around the world and help get materials to the third world countries that we will be helping as soon as practically possible. The air force with their large transports will be used to take heavy machinery to these countries and it will all be free.

"The army and its engineers will help improve our roads, build new factories and generally help where they are needed throughout the world, again especially in third world countries, where there will be so much work to do and teach those people how to operate the machinery to help feed themselves.

"The last thing we, our country needs to do is change our faith and the design of our churches which I hope God will do for us, at least that is what Jean has explained to me. So now I am asking you my people; listen to Mary, First Lady Rocc and their disciples and make up your own minds. There is still plenty of work to do in rebuilding houses and roads and improving the way we live. I have opened our all boarders to allow people in and out whenever they like. Those people coming in will only have to agree to one thing, at least listen to our new faith and at least once while they are here, visit a new church to pray to the new God.

"I am also having talks with car manufacturers around the world about bringing their companies here to make their new cars and bring our country up to date. You will also from now on have free access throughout our entire country to Wi-Fi and the full internet. As I learn of new things that will help us, I will implement them into our society. So, all I ask now is that for a few moments you all listen to Mary of the new church of the world."

Mary shook his hand, kissed his cheek each side and walked up to the batch of microphones.

"Thank you President Dmitriev for your very warm welcome and the kind words you have said. The world is changing on an unbelievable scale; at an unbelievable rate: I am here to answer your questions and help bring the new church to everyone. We will be here for some time; although I will have to leave shortly, but as President Dmitriev made such an impact on me when we met at Buckingham Palace, I wanted to come here and officially thank him for welcoming us to your country and to welcome you into my Mother's Church. First Lady Rocc and my disciples will make many new disciples of Russia to follow in our work.

"As you all know, I have built a new spaceport and shipyards in Russia and you can help build the spaceships. There are thousands of jobs there for you to do and you will be very well paid both in money and material objects. The other thing I have not told anyone before, some of the people who help build the Arks will travel far into space on them, they will be asked to stay with the Ark and work on it as part of the crew when they depart from Earth. The people, who wish to go, will also be allowed to take their families with them. I hope, with President Dmitriev's help, to build many more ships here and each time some of the ship builders will be asked to join the ship they help build and if you did not wish to travel on the first or second ship, you might wish to travel on another, and anyone can request to go. There is plenty of room and plenty of jobs for people to do while in flight to the distant planets and stars.

"Most of all, I am here to tell you all about the new church and my Mother, the real true God and also, to tell you a little of my Father Lucifer, who, from this day forward, despite Him taking control of the Earth and you people here, I have feelings for Him and I would like the people to respect Him and refer to Him by His name Lucifer," Mary said and then paused as she saw a woman look at her and raise her hand to ask a question.

"Excuse me for asking a dumb question Mary, but if your father is called Lucifer, you refer to your mother as God, what is your mother's name; as we have never been told and you have not brought the subject up?" the woman asked from the front of the crowd.

The entire crowd went silent, even the birds, which had been signing in the background and flying overhead settled down and stopped chirping. The plane, which was moving around getting ready to taxi to the main runway, appeared to drop its engines revs and the noise of the aircraft died away. President Dmitriev looked down at the woman and questioned without speaking, why she had asked the question, but as it was asked, he admitted he too would like to know the answer. He thought even more and decided everyone in the world would like to know the answer to that question. What was God's first name?

Mary looked down at the woman and quickly read her thoughts, she was not simple, she was quite intelligent, but she wanted to feel closer to God and therefore, as an act of decency and good wishes, wanted to know the person she was praying too. All she wanted to do was know her name, so that she could at times of great need, call her name and ask her a direct question or request her help, even if it were not given. There was no other reason for her question; it was one of honesty and trying to be polite to God.

"Valentinia," she said speaking the woman's first name. "You should not fear me for asking this question, it is one that I must answer, had I been asked it on the day I arrived on Earth I would have answered it then. Actually, believe it or not, you have been addressing my Mother by her name all your lives. Her name is Godisious Alexandria Morthania Jozanda Dalaxia; My Father is called Luciferian Candellioux Augantain Dalaxia. We are as a species, called Primes or Utopians and there are a lot of us who create various universes and yours is one of many universes that our species care for.

"Her name for short is God, and My Father's is Lucifer. We are a grand species who can alter matter with our minds and blend it with our hands then stabilise it and transform it. All our names are quite long and everyone shortens them like you do with your names. So when you say God, you are correct in calling Her that. I know you have been praying to God, but Lucifer has taken my Mother's name in vain and taken control of this planet and for want of a better word, zone, as that is what we call these places. I can tell you no more at the moment, but hopefully in the future, when people come to the new church, my Mother may be able to explain more about this zone to you, but this is for Her to decide, not me."

"Thank you Mary for explaining this, somehow, I don't know how, but I do understand what you have just said and it makes perfect sense to me," Valentinia said, surprised by her own reply.

"That is because I have allowed everyone to understand my answer and hopefully this will bring you closer to the new church," Mary explained.

After a few more words of thanks and where Mary and her disciples would start their missionary work, President Dmitriev invited Mary's group to a feast where every person who was in the airport was invited to join them. It was very clear to Mary, that President Dmitriev was a changed man and had opened his country to her. He was trying to get his people to believe in God and the new church and was not standing in her way.

Three hours later, a large coach arrived and First Lady Rocc and her disciples boarded it and drove off to their first destination. Mary joined them for the first part of the trip and spoke to her disciples again before leaving them and getting on with her own job. When Mary left, Master Yegor, a new disciple made during the first hour of them arriving in Russia, moved from his seat and asked to speak with First Lady Rocc.

"Please sit down, how can I help you?" Jean asked.

"I apologise for asking this question, but why did Mary join us and take over your introduction if she has just left us?" he asked.

"Ah, that is a very good question and one that I can answer. When we were in Buckingham Palace, everyone was against us until President Dmitriev came forward and shook her hand and spoke up for us. It was right for Mary to return the favour and be here to introduce us to the Russian people. Also, President Dmitriev wanted to speak with Mary privately and this was an opportune time for that to happen."

"I didn't see them talk together, they were with people all the time," Master Yegor replied.

"It only appeared that way. During the time we were eating and talking to other people, Mary took hold of President Dmitriev's hand and took him to a different time zone; there they spoke in comfort for over ten hours, discussing many things and what Mary wanted for Russia. President Dmitriev also wanted some jobs done, which have already been completed and as we speak, more work is being carried out by the angels that are helping us to build a better world in which to live."

"What sort of things did he want Mary to do?"

"Simple things to help his people; every railway has been upgraded to the latest trains and carriages, new lines have been added across the country and through northern Europe joining Russian countries together. Two more large spaceports have been made complete with the infrastructure to support them, but there was no way the people could get to them, so Mary has taken care of this. She has also upgraded every road in the country, laid new power lines, underground pipes and sewage pipes, drains, water pipes and gas lines. Mary has updated every power station; gas drilling platform and made all the electric and gas in Russia free for its people to use.

"She has also built the very latest hospitals and medical centres all over Russia and the countries in northern Europe. All this was done during their discussion in the Time Zone. There was also another reason she was here, to answer Valentinia's question, which she thought should be answered because this was the only person in the world who thought of asking this one specific question, a question I should have asked her when she arrived two weeks ago. Mary was hoping, by answering this question, her Mother would become better known to the people of Earth," First Lady Rocc explained.

"Ahh, now I understand, thank you so much for explaining this to me. I would just like to say, thank you for picking me to join you, it has always been a great wish of mine to do good and join the church, I just didn't know which church to join, but now I do, and it all makes sense to me."

"I know; that is why you are here with us and you have been made a disciple, not only a disciple, but the first disciple in Russia. Your name will go down in the church archives for all history, and it will become a very well used name, many mother's will name their children after you," First Lady Rocc replied.

"Then I am honoured," Master Yegor replied.

Over the following week, the disciples split up travelling around Russia bringing the new faith to everyone they met. They answered questions, even did some minor healing of burns and broken bones. Mainly they answered questions and spoke passages from the new bible, telling people where they had gone wrong and how to pray to the new God. As they came to a church, two disciples entered it and spoke with the local vicar or minister.

As soon as word got round that a disciple was in their church, the people flocked to the church and the disciples waited for more people to come then started to give their sermons and spread the word of God. As the church was filled and over flowing with people, the ground shook and God changed the church to the new church style and altered the inside around the people so they saw the miracle happening and were moved to believe in the new God.

As more and more people changed their belief, God was becoming stronger and stronger and so too was Mary. As more people believed in her, the aura around her grew stronger and quite often, Mary was seen to glow by those people who had second sight. God was also becoming more powerful and Lucifer was losing his strength, something Mary knew would happen as time went on.

People in Russia were now flocking to the spaceports to help work on building the massive Arks and smaller spaceships. There was one spaceport that built nothing but small research vessels, fast and slim, carrying four people who could get in between asteroids and even land to collect samples of rock or other substances. The factories at one of the smaller towns close to the main spaceyard made the spacesuits that the people would wear once in space and the thousands of new uniforms for the space research teams.

Five hundred of the new spacesuits were sent to NASA and the current Russian Space Agencies for their own astronauts to use to get into space on their current rockets to get to the ISS (International Space Station), which was still operating.

The same spaceport also made single and two team support craft that carried powerful lasers and cameras to film God's wonderful creations. The laser could also be used to cut off rock from minor asteroids for further examination or destroy comets or asteroids that were likely to collide with a planet and kill vast numbers of people.

In Kazakhstan, where the Russians sent their large rockets into space, sending up the astronauts to the ISS, they were now making the large shuttles for the huge Arks. For allowing them to use their facilities, which Mary and God had enlarged, the first ten shuttles they built, Mary allowed the current space agencies to use these to take the astronauts up and down to the ISS. She also set a team of designers the task of building a massive Super Space Station Wheel. It would be up to the Americans and Japanese to build the parts for this and they had only been given three months to complete the job. To accomplish this, Mary gave them anti-gravity tools to help move the parts around and she gave them three sections of space station to be fitted together and then they would have to complete the insides and load all the equipment into it. Wondering how they would get into space, all Mary told them was to have faith. If they didn't, it would stay on the ground until they did.

To help the people build the space station and help in their beliefs and change the way they prayed to God, four new disciples were made from the workforce who spoke to the workers and took their daily morning and evening prayer sessions and talked about the new bible and new church. A new church had been made by God at each spaceport just for the workers. At first it was only used by one or two people but as the days progressed, more and more of the workers and their families started to fill the pews morning and night until every day, the church was filled and overflowing, so much so that the church was enlarged without anyone realising what had been done.

As the space station builders believed more in the new church and listened to the disciples speaking to them, more builders and their families turned to God, and as they did, so their work on the new space station seemed to go faster. In less than four days, when seventy five per cent of the workers and their families had changed their belief to the new church, every one of those people worked or moved forward easily and every part seemed to fit, no matter how complicated the parts were to get into their position.

As for those people who had not yet joined the new church, they were still finding their jobs hard, and each part they had to fit were more and more difficult, so much so that often they gave up, sat down in frustration and called for another engineer or electrician to help them. When they saw how easy they made the job look, how they got the parts to fit and the speed in which they moved, the other people started to believe in what they were doing and also started to attend church twice a day and say their prayers and talk to God before they went to bed, day or night as they worked around the clock.

Some even just lay in their beds for a few minutes, saying their final daily prayers, then quickly fell into a deep sleep and woke in the morning fully refreshed, ready to face their daily chores of church, breakfast, prayers for their meals, work and home for another ten minute prayer session at the local church and a chance to thank God for how their day's work had gone.

In three months the new space station was completed, and a further week was taken up loading it with food supplies, experiments, bedding and clothes. New spacesuits were also packed into the space station and everything was given its final checks and tests before head disciple Master Paul Constantine blessed the Space Wheel called Genesis, after the first book in the old bible, but now meaning First Space Wheel, which was standing upright, a kilometre high and half a kilometre wide with spokes and walkways to each section of the station. There were a hundred access ports to the station and two thousand escape pods.

Genesis would hold three thousand people, scientists, engineers, technicians, medical personnel, children and caterers, every job had to be filled and many of the jobs for the maintenance and working of the space station went to a quarter of the people who helped construct it.

Two Universities were given a hundred places each for students and professors to work in space carrying out experiments that would help the people on Earth. There were also four small churches on the station and at all times, two disciples would be there. They would not have to remain on the station all the time, but work in three month rotations, taking their families with them if they wished to go.

The people of Earth were now realising their future was no longer on Earth, but in deep space and Mary was preparing the world's population for long term space travel. Already the new space shuttles, needing no heavy rockets to get them into orbit, were ferrying NASA and worldwide astronauts to the ISS. Four space shuttles were always travelling from the Earth to the Moon taking machinery and living accommodation there four times a day. The shuttles took an hour load, an hour to get from the surface of Earth to the Moon and two hours to unload the equipment and get it positioned where it needed to go.

After two weeks of sending the first materials to the Moon, a crew of two hundred engineers were taken to the Moon to work there. Two new shuttles were allocated to MARS ONE who were sending more and more materials and new homes to Mars, with the trip to Mars taking a very long six hours by the large shuttles, which had been made for this special purpose. The thirty people on Mars were now joined by another hundred engineers who were putting together the Martian homes for the next settlers.

As Mary brought in more female designers, they started their designs for two world terraforming machines. The Japanese and Indian engineers where given the task of making these enormous machines and once again, two disciples were taken from the group of workers to help bring the church to the work place. This time, the workers were very interested in the church and as soon as the church appeared on the work site, God had to enlarge it overnight as when the two disciples entered the small door at the rear of the church, to their amazement, every worker was already inside waiting for the service to begin. They were already on their knees, praying to God to help them in their daily work, and one of the site supervisors, was standing at the front leading the simple prayers.

Disciples Anne and Rosemary, watched in amazement as the man, Seijo Kaba glanced their way and finished his simple prayer. He bowed his head low, and started to tremble before the disciples as they moved slowly towards him in silence.

"Please, so very sorry for what I have done. No one was here and some of the workers wanted to pray, I so sorry to take over your position. So very sorry," he said, almost in tears.

"Seijo Kaba, please do not be sorry, we applaud you for what you have done. God will be very pleased you have taken the initiative to start the prayer session; it is us who are sorry for arriving late. We did not expect the church to be this full so early in the morning," Anne said.

"Only workers here now, families come later after they have made breakfast and before sending children to school. Once again I offer many apologies for starting without you."

Anne placed her hands on his head; she said a prayer and blessed him and the people inside the church. The Church shuddered for a minute as it grew larger to accommodate more people and the Holy Spirit fell over the workers. That day, everything the workers touched went together perfectly. The nuts, bolts and screws all fitted flawlessly and they did more in the first day after prayer than triple the workforce had achieved on the other space station.

As soon as Anne lifted her hands from his head, he could see much clearer how he felt about God, he could speak every language on Earth and many more that he would need when he went into space. He had a full understanding of what the machine he was building had to accomplish and how it operated and what had to be achieved to get the work done. He looked up and pointed to two men, calling them to him as if instructed by an unknown force. The two men, trembling, shuffled their way forward, their hands together before them.

When they got to Anne and Rosemary, they again bowed their heads low.

Rosemary placed her hands on one of the men and Anne did the same to other man. A moment later both men looked up and smiling, understood what gifts they had been given. They turned to face the workers before them and spoke very quickly in their language and blessed everyone there.

Anne and Rosemary took the remainder of the service and later, the three new disciples joined them in the church when they were instructed as to how they could help pass on the word of God through stories in their lunch breaks, which these workers seemed very pleased to want to hear. They were hungry for the new church, to learn more about it and how God would change their world and help take them to the stars.

The work on the terraforming machines went extremely well, and everyone who came to the site; visited the new church first to thank and pray to God for Her help with the designs and getting the machines made. As many of the workers and engineers where Japanese, they all seemed to have more respect for the church and soon, Seijo Kaba approached Disciples Rosemary and Anne.

"Thank you for seeing me," he said now in perfect English.

"What can we do for you?" Anne asked.

"A new engineer has arrived at the site and has come to our church and after morning prayers; he has come to me and asked for some help. Emperor Chan has spoken to Chief Engineer Tekkan Itoh, and has given him a message as he was coming here, Emperor Chan has asked through Tekkan Itoh to request a meeting with you but Tekkan Itoh is too afraid to speak with you and so has asked me to pass on the message from our Emperor."

"What is the message and how can we help, you know we are here to help you and please do not fear us, there is no need," Rosemary explained.

"Emperor Chan has heard of your church and has requested that you send more disciples to Japan and help spread the new word of God. He would also like to meet you and talk about the Mother God as he wishes to know how is best to pray to Her. He is greatly humble to hear of the new God; can you help us with this problem?"

"I have just passed your request to Mary, the daughter of God and she is sending First Lady Rocc to meet with Emperor Chan later today. She will take with her a hundred disciples who will help spread the word of God to all Japanese people if it be the will of Emperor Chan. Can you ask Tekkan Itoh if this will be acceptable?"

"I will go immediately and ask him, he is right outside this church."

"Then bring him in, I will speak to him in his own language of course," Rosemary replied.

Within ten minutes Tekkan Itoh had contacted Emperor Chan's secretary and gave him the good news and First Lady Rocc was getting her disciples ready to leave. Within the hour, they were sat talking with Emperor Chan and realising how the Japanese people were so eager to learn about the new church. First Lady Rocc was passing everything Emperor Chan said direct to Mary and suggesting that the word of God had spread far faster than they ever thought it would. First Lady Rocc asked for a further three hundred disciples to cover Japan and change the entire county to the new church with Emperor Chan's permission within the week.

With Mary's blessing, the extra disciples were sent to Japan and what made life much easier was that they could all speak fluent Japanese and every other language on Earth and throughout the Cosmos. As First Lady Rocc had said, with the extra disciples and the five hundred disciples they had made in Japan, they converted every church and almost everyone in Japan to the new church. It was only one area that was very remote that had not been changed and first Lady Rocc went there with five disciples to make a new church and pass on the good news about God.

The news of the new changes was also being heard in China and the Chinese leader President Changpu Chang also wished to hear more of this new church. Mary had brought the north and south of Ireland together and changed their faith, a very difficult job indeed, especially with the problems with religion. The other job Mary had done was bring the whole of Ireland, Scotland, Wales and the small Islands around the UK together and ruled by one person, King William V; which would help her bigger plan.

While Mary was working with First Father Langley and six hundred disciples in China, Mary was helping Catherine Patel, who was now a Full Disciple and the Director of Concord Aid in Israel. She now had under her a staff of two thousand people from around the world and many of them from Israel.

***

Concord had been made much larger and now had another fifteen large buildings to help organise world help for those people still in need. They also handled all the incoming machinery and equipment being sent to third world countries from the UK, Russia, France, Spain and America. These countries were also sending engineers and spares for the machines for when they broke down. Many of the machines and tractors had been changed from diesel to electric motors, which were charged by solar power batteries packs placed all over the body of the machines. As they worked in the hot sunny climates, the machines worked continually throughout the day and they didn't have to move fast, as harvesting and planting was always a slow process and usually done at walking speed.

Every engineer and person who entered the work programme had to pass through Concord and sixth Mother Catherine Patel gave the power of multi language to everyone as they would be helping people in different countries in the fields. Many of these people were also made disciples, and during their meal breaks, got groups of workers together and talked about the new bible passing on the word of the new God. When they were in the right areas, using their telepathic abilities, they contacted Catherine Patel and requested a new church be built and whenever Catherine Patel prayed to God for a church, her prayer was always answered and it was placed in the position where the disciple had decided it would best be used. So far throughout Nigeria, Libya, Western Sahara, Jordan, Yeman and Djibouti, over two hundred churches had been built and the disciples had made another eight hundred and fifty disciples to help the people of these countries learn to pray and pass on the word of God.

Not one of the disciples was wrongly chosen: those who picked their replacements to continue their work as they moved on to another village were all passed by God, and She told Her disciples if they were good and believed in Her.

Mary was constantly helping Catherine Patel, giving her guidance and telling her what she wanted her to do to pass on the teachings of God and Catherine passed on Mary's commands at every morning meeting, after she had led the workers in their morning prayers at 06.00.

***

Nobody seemed to understand the meaning for the church to start their main workers prayer sessions at 06:00 and again for children at 09:00. What was actually happening as more and more people joined the new church, at each prayer session, all the voices were amplified as at the top of the church, was an acoustic amplifier to increase the sound of the prayers and thoughts of people when they prayed in silence, which was included in every prayer session. The thoughts and voices were amplified which gave God more power and took the power which Lucifer had been having away from him so He was becoming weaker and weaker and his power of the people of Earth was draining at an incredible speed.

The other thing that the amplifier did was send their thoughts out into space to four species that were on the verge of contacting the people of Earth to ask them to join a Federation of Planets that would hopefully bring peace throughout the Cosmos. They were still many years away from contacting the people of Earth, but many scout ships had visited the planet Earth to see if they were ready to join them. With the new prayers, the species would now get the message the people of Earth were changing on a massive scale and getting ready to launch their people into space.

There were still a lot of men and women who prayed the old way knowing know they were praying to Lucifer, and continued to worship Him. Mark was still so weak he had not been able to help his Father in His work to distract the people from the True God, but his mind was not in his Father's work any longer, he only wished to be with his Mother.

When the children and mothers prayed at 09:00, the mothers were now invited into the schools to the prayer sessions, and those who couldn't attend a church, took a few minutes in work to fill the work churches, a small room where the staff and one of the managers, who was a church disciple, led everyone in prayer for five minutes. Without anyone knowing, God had included at the top of the room a smaller amplifier which again amplified the thought of prayer to Her.

After only three weeks of the new church being formed, everyone throughout the world, were getting used to the new prayer times and even in shops, people were used to the tills closing for seven minutes at prayer times as people made their way to the prayer rooms and back to their work stations. Many people, on hearing the new prayer chimes, a tinkling cascading chime; would know when the chimes stopped the prayers would start.

In the shops, many customers joined together in a group and said silent prayers or asked a member of staff to lead them in prayer. If there was room, some shoppers were allowed into the prayer rooms to join the staff in their prayers. When the chimes started again, it let everyone know that the prayer session had come to an end; but they did not have to stop praying then; if wished, they could continue singing songs and praising God.

Everyone had their prayer alarms set on their watches and phones or alarm clocks and everyone was now getting used to the new prayer times and with the help of Mary and God, feeling a lot better as the air had been cleaned and with nobody now suffering from dementure or memory loss, it was a simple way of saying thank you to God.

Mary was still thinking of another problem she wanted to solve and knew the time was soon approaching when she would start her biggest reform of all. Not even her Mother knew what she was planning and having spoke an eternity with Her, knew what her Mother wanted and had always missed. Mary was hoping to get the Earth right first, and then the other problems would fall into place.

Once a week, Mary and first Lady Rocc made a quick tour of every spaceport seeing how the work was progressing on the new Arks and shuttles. Eight shuttles and two very large shuttles had been given to NASA, MARS ONE and the Russian Space Authority. Four more shuttles, as soon as their final checks were completed, were being given to India, China, Japan and Europe. The UK's space plane Skylon, which was now in its final stages of being built, was made twice as large. Within two weeks it was flying into orbit and taking supplies to the ISS. Sir Richard Branson's White Knight space craft was also made four times larger and given more efficient engines to get it off the ground without the need for its Mother Plane. There were still eight of these craft used to take people into space and show the first public astronauts what was happening at the space station and to look down upon the Earth and take photographs of what they saw.

Six months later, four of the first Space Arks were nearing completion. Work had progressed very fast, and with help from a number of angels, the outer and inner hulls had been completed and the engines had been installed. When the bridge and the engine room were completed, the engines were tested for the first time. They were started and despite their size, there was hardly any sound coming from outside the ship.

Captain Amelia Carter of the Dauntulous and her bridge crew monitored all the instruments as the Chief Engineer, Lyusya Mikhailov, and her staff started the main engines and checked the ship's generators and force fields.

All the workers were well away from the ship when the test was started, four disciples were outside the ship just in case they were needed and two disciples, one on the bridge and one in the engine room were there to help if they were needed. The disciples told the crew they were only there to see how the ship operated and had full faith in their work. This was one job which had to be done and it would better to pass this test now while the engineers had access to all the ducts and cables. Once the rooms were completed, many of the cable ducts would be covered and it would require draw wires inserted to pull more cables and instrument wires through the ducts if cables failed or more cables were needed to operate machines and equipment.

The ship commenced its first test at 09:30, thirty minutes after the morning prayers which everyone on the site had said an extra prayer for the safety of the crew and their prayers that their work was right and the ship moved.

As the engines came online, a number of loose objects started to lift into the air as the anti-gravity engines lifted the Ark Dauntulous, three kilometres long, one and one and a half kilometres wide and one and three quarter kilometres high effortlessly from its holding dock. The eight hundred support struts were retracted into the hull and the hulls sealed. A number of engineers sent in automatic cameras and detectors to ensure the hull was properly sealed when the struts were retracted into the ship and the outer hull doors closed. A second door closed inside the ship ensuring the outer hull was secure in this area and air would not escape from the ship.

Once the checks were completed, the ship lifted a further hundred feet into the air and held its position for thirty minutes while tests were carried out on the engines and generators that supplied internal power to the ship. While the ship was in the air, four large special cannons fired huge rocks at the ship when the force field was activated. The rocks never got to the hull, each one was destroyed by the force field; breaking the rocks into millions of small pieces.

Next, a tank was brought in and fired ten metal piercing shells at the hull, but again not one shell reached the hull, it was destroyed under this test, a hundred feet from the ship. The tanks fired at the ship from different locations and at different parts of the ship. They had to know there were no problems with the force fields. Two lasers were also checked from the bridge as two giant rocks weighing over a hundred tonnes each were fired at the ship. The lasers hit the rocks and broke them to smithereens which were again dispersed upon colliding with the force field.

To further assist with testing the ship's force field, two jet fighters were given missiles and sent into the air to fire at the superstructure. The planes also fired their cannon, which Mary had allowed to be given back to the Air Force for these specific tests only. The planes also carried four bombs each, which were dropped onto the ship, but there were disciples in position in case the bombs went astray and could be stopped from hurting anyone should they miss their target.

Eight hours later, with the tests completed, the support struts were deployed and the Ark lowered on onto its struts which now took the full weight of the ship. Once the leg struts had taken the full weight of the ship, it lifted a short distance into the air again and main dock supports for the ship were moved back into position and the ship lowered on the dock struts. This allowed final checks to be made on the ship's main struts and the final coat of powder paint could be applied to the exterior of the ship.

At last, when the engines were shut down, the main electric supplies were attached to the ships' hull plugs so it could now run from shore power. The mains water and sewage pipes were also connected so the workers could use the toilets aboard the ship and the four restaurants that were now fully operational aboard the ship, could feed workers and crew.

With these tests cleared, the people throughout the world gave a massive cheer on seeing the huge ship lift under its own power for the first time. Never before had anyone ever dreamed that a ship of this size could get into the air.

Two days later, the huge orbiting wheel, a kilometre in diameter and half a kilometre wide, was completed and under its own power, with a crew of two thousand people aboard, all their food and supplies, eight shuttles and sixteen support ships, all the escape pods, with the people around the world watching, lifted off from its base in Houston and ascended into the blue sky.

It rose very slowly at first as the wheel, upright, stabilised and moved higher and higher into the air. When the space wheel was two kilometres high; being followed by a number of air force fighters that had been converted to camera planes, followed the wheel to their ceiling height. Six small two man interceptors from the Russian Baikonur spaceport followed the space wheel on the rest of its voyage into orbit 380 kilometres above the Earth.

The entire trip into orbit took a very short twelve minutes and then the space wheel started its own generators and gravitation generators to give the crew Earth's gravity at sea level. The outer ring although made of a clear form of plastic, was covered in solar power collectors to help power the space wheel. A large microwave emitter was deployed from the centre of the wheel which sent excess energy from the wheel, down to Earth in microwaves and helped power North Africa,

Twenty large shuttle docking tubes deployed to their permanent holding stations. These would be extended when a shuttle wanted to attach to the Space Wheel called Genesis, being the first of many space wheels to be made. From the central hub, a much larger docking tube could be extended to dock with the Arks and any other large ships which would be built over the coming years. This would allow the transfer of goods and people to the large ships if needed.

Two large rescue shuttles were also under construction, able to take five hundred passengers each plus a huge cargo if Genesis was ever in danger of being hit by large meteors or a sun burst of high radiation or gamma radiation. Genesis had its own force field and laser defence system and was also able to raise or lower its position in orbit.

Now the first Space Wheel had been completed a second Space Wheel was under construction in Canada and replacing the empty work area of the space wheel in Houston, was going to be another Ark, this time five kilometres long, two point six kilometres wide and two point five kilometres high. This would be the second of the larger ships to be constructed, the first ship this size was still under construction in Australia, while in India, the fourth smaller Ark called Enterprize; was nearing its completion. Once again India, like Japan and China was eager for the new church which helped to stabilise the countries.

As soon as one country was turned to the new church, the large group of disciples were moved to the next country and they moved through it as fast as they could, making new disciples as they passed through and leaving one or two senior disciples to lead the new ones, who were quickly moving up the ladder. The more people they got to change and remain changed to the new church, the higher up the ladder they moved and the more responsibilities they were given by Mary.

As First Lady Rocc and the remainder of the first disciples were travelling around the world, Mary realised she would need more people to assist the higher disciples which now numbered over thirty thousand.

These disciples were being put in charge of churches and leading the prayer sessions, travelling around schools, colleges, shops and work places helping out wherever they could. The hardest religion to change was still Islam, which was not cooperating with Mary's group at all and still praying to Lucifer. It was Iranian Rukia Jarfari, who after speaking with disciple Jeffery Carr, burst into laughter and after apologising to Rukia, he passed her idea straight to Mary.

Mary appeared two minutes later in Iran, inside the home of Rukia Jarfari and spoke at great length with her. After an hour, Mary requested a meeting with the local Imam who was Mahmob and despite him not really wishing to speak to her, knowing she was thought to be the daughter of God, he agreed but he would not meet her in his Mosque. They met at a local market where they drank tea and talked.

"I can fully understand that your religion is debating the sex of God, this was something you were not prepared for and as I explained, you have been throughout time worshiping Lucifer.

You have seen the many miracles I have made and the miracles my Mother has made. Wherever the people wish a new church to be built, it is built in a few minutes with the power and heating all free. I believe at the moment all your Mosques are connected to the main electricity supplies which you have to pay for?" Mary asked.

"That is correct but we do not mind this, it is something we have to do," he replied.

"Do you think there is any chance at all that your church leaders will change their minds and change to the new church?" Mary asked.

"Being honest, I think not, I would like to have said yes, and secretly, seeing what you have done, I believe in you and the new female God, it is just that my superiors will not allow women to lead the congregation, they do think it right and they do not wish to be governed by women. It is their belief and opinion that women are the lowest people on Earth and should be treated so."

"Surely not everyone in your church thinks this way?"

"The problem is it is the elders of the church and their minds will not be swayed."

"They will have the surprise of their life when they die and meet their true God."

"No doubt they will, but while they are here on Earth, they will treat women as they see fit. That is why so many of them have to cover their faces. Inside they do want wish to see them looking at the men because they know exactly what the women think of them."

"On looking around I have noticed you have conventional churches here, are other churches allowed in Muslim countries now?"

"It was one of the concessions we have had to make. When our people moved to the USA, the UK and other western countries, Russia included, we wanted and demanded to build our mosques in their countries so we could say our prayers. The Governments turned around and told our leaders, if you want to have your churches in our countries, then you in turn must allow our churches in your countries and allow us to preach to our God in your country. If you do not agree, then do not come here, return to your own countries; so the church leaders were given no option but agree. You can build your own churches in Iraq and pray in them if you so wish. You are also allowed to try and change other people's beliefs and we are allowed to do the same in your countries."

"Then Mahmob, that is what I will do," Mary said strongly

"Do what?" he asked.

"I will build and place my own churches here and fill them with women and make more women disciples, I will not force the men to change, or ask them, I will just talk to the women, you just said it was allowed."

"But you are the daughter of God."

"And you should listen to me, not my Father."

"But . . ." He could see he would not win the argument, not that it was an argument and Mary was speaking to him fluently in his language, he did not have to try and speak the little English he knew.

"I see language is a concern to you."

"It is something I have tried to master, but alas, I have so much to do, it is hard to find the time to learn more languages."

"Please kneel before me."

"What!"

"Kneel down, I am the daughter of God, nobody will see you I promise."

He looked around then got on his knees, Mary placed her hands on his head and moment later he felt something very strange happen to him and a when he opened his eyes, he found he no longer needed his glasses, he could hear people talking on the next table and knew they were from Vietnam, yet he could understand every word they were saying and they were talking in their own language. His eyes beheld four Chinese people at least a hundred metres away and when he looked at them and concentrated, again he could understand every word they said despite the fact they were speaking in their own language.

"Thank you, this is marvellous, I know what you are thinking now and it will work I am sure, you have converted your first Muslim to your new church. Go ahead and build your churches for the women, I will help you all I can and I think there may be a way you can do much more."

"Please tell me what you have in mind," Mary said and with two more drinks before them, listened with intrigue to what he had to say.

The following day, with fifty new churches now in Iraq, the churches were filled with women and they were taking the prayer session. Not one woman was attending the mosques and when in the new church, they were all allowed to remove their headdresses. At midday, as agreed, Mary met with twenty Elders who ran the Mosques in Iraq. She spelled out Mahmob's idea and showing what he meant, she picked one man and had him kneel before her. They had all wished to be picked but quickly reading their minds, Mary picked the man who was liked the least and gave him the power of multi language.

"When the Space Arks are ready, those with multi language abilities will be allowed, if they so desire, to take their families and travel far into space and meet many new species you have never before imagined. You will be travelling for many years and for this, I will give you the gift of extended life for a minimum of a thousand years."

"Why did you not pick me?" Naadir one of the elders asked.

"No reason I closed my eyes and asked God to choose for me. Do you wish to have this gift?" Mary asked him.

"Why yes, it would help so much in preaching and I would like to travel to the stars."

"I could not allow you to travel to the stars if you only wish to preach about Lucifer; that is not what I am doing."

"I will change my beliefs, I will allow everyone to change the way they wish to believe in God and allow the women to teach us how to pray; in return, I would like to travel to the stars," Naadir replied.

"You will have to promise you will change your ways; cut your hair, stop women from wearing the headscarves and admit the new God is female," Mary explained.

"I . . . I will do so and I will talk in one of your churches to the women and explain how we are going to change and they will no longer have to wear their headscarves in the streets. This will all start from today when I enter your church at 18.00 for the first of the evening prayers if you will allow me."

"Then so be it, all of you, please kneel before me," Mary said and when they were all knelt, she blessed them as a group and gave them the gift of multilanguage and offered each person a position on one of her ships to travel to another galaxy.

Once they had been given their gifts, they were like changed men, they went out and talked to every man they encountered, passing the new word of God and entering a Mosque each, told their people what was expected of them and that from this time forward, they would be praying to God, a female God and they would have to change their opinion of women and allow them to go around as western women do. A few men shouted no, but the new Elders spoke with loud voices and powerful words, and very soon, the people inside the Mosques found they were now in the new church, which was a different shape and here, they prayed to another God, a female God.

When the men turned to the new faith, they felt a change inside them; each man felt better, calmer, no longer was he angry, and no longer did he want his wife and female children to cover themselves with black cloth; they would now be free to be seen and respected like other women throughout the world. They would also be allowed to take the service in the new churches and they would follow the new God and pray in the new church.

Within the following week, with the twenty Elders meeting all the other Church Elders, they had managed to make their church leaders see the way of the new church and change their way of praying to God. They had even agreed that women were equal to the men and inside the church, they were higher than them and that as God was a female, She wished her female children to be in charge of the church, it was for Her; only natural and after more discussions with Mary and First Lady Rocc, the men agreed for women to be in charge of the churches in all Muslim countries.

Mary had agreed and put in writing that the twenty Elders she first spoke to would be allowed to travel into deep space on the fifth Ark. That is when she told them they would be with the first people to travel through a Black Hole and arrive at another galaxy, the Great Spiral Galaxy of Andromeda. When they heard this, they were thrilled at their new quest and agreed they would do everything possible to help change all Muslim countries to the new church.

It took the Elders and three hundred new disciples four weeks to speak to all the leaders of Mosques and get them to change their views and covert from the Muslim religion to that of the new church. They also had to come to terms with allowing women to take over and lead the church in prayers.

At first the other Elders were dubious, but after they were told some of them would be allowed to go into space, it was surprising how they came to change their minds. Many of the Elders would not be travelling into deep space, some would go to Mars and two other Moons in the solar system, others would move on to new planets close to their system, but they would get to fly in space and they would get to do what they most wanted, spread the word of the true God.

When first Lady Rocc and Mary gave some of the Elders and church leaders their new gifts, they felt so different, some men said they no longer needed to go into space, they just wanted to help the people of Earth come to know God better. Above all, the biggest change was that the Muslim women would no longer have to cover their faces, they would be allowed to show their faces and their bodies if they so wished. God did not want women to cover up their physical shape; she wanted them to love their bodies and believe in themselves. There was no reason to cover their body, not when God ruled the Earth and the women would no longer be raped or attacked by men. If they were, the man knew he would suffer for the rest of his life and be changed to a woman and have green hair that he could never change.

He would be marked by God, and everyone would know what he had done and be punished for the rest of his life. It was like a death sentence, people did not kill because they knew they would also die if caught in many States of the USA.

The new church was finally recognised throughout the world and there were only one or places that did not have new churches. People were starting to listen to the disciples that spoke over the radio, many of which had been dropped from aeroplanes over the jungles of South America and Brazil. As people listened, the word of God was spread throughout the Earth. Mobile phones which had also been dropped from the aircraft needed no battery to speak to the disciples and ask for help and understanding or speak to the people at Concord Aid. As soon as the people gave them their location, a disciple was sent to the people who were usually in small tribes, and the disciple helped to convert the people to the new God and even gave them a church, which had electricity, running water and air conditioning. It also had a television, computer and a proper phone system where they could speak to other people around the world.

Two weeks later the disciples moved slowly and passionately through the Amazon, Brazil and the farthest parts of Mexico where people had been missed before on an earlier pass. Thirty more churches had been built, all small; but all important for God to know Her children had the opportunity to pray to Her in Her Churches and take the credit away from Lucifer.

While all this was happening, Mary was still making her plans for her biggest conversion to the new church. The four large Space Arks had been checked for flight and their generators and force fields tested. All that had to be done was fit out the inside and that job was well underway. Most of the duct covers had been put in position and there were thousands of people working on each Ark and the fifth Ark Explorer; had just passed its lift off, generators and force field trials. The four smaller Arks were getting further along as more and more people helped to complete the ships, being paid very well in materials, health and money, but above all, the people knew many of them would be needed on the Arks to help keep them moving through space and they were big enough to accommodate over fifty thousand people comfortably. There were more than that working on the inside of the ship, let alone the outside and running cables between the hulls and beneath the decks.

CHAPTER 20

Ten months after Mary had arrived on Earth the whole planet had changed. The skies were much bluer, the clouds, except when raining, were clear white, the seasons had returned to their former months and for the UK, when it was winter, the river Thames froze and people, for the first time in over a hundred years, had fairs on the frozen river, until the ice breakers came through and broke the ice up for shipping to get through, but it was a wondrous event and the people were very happy.

Around the world; the huge deserts were no longer deserts, but irrigated farmland growing maize, corn, wheat and every vegetable there was. Huge pumping stations converted the sea to pure drinking water and this was pumped through massive pipes, all supplied by God including the pumping stations to irrigate the deserts.

Those people who were dying of thirst and hunger were now starting to farm for themselves, feeding their own people and being taught how to farm by farmers from around the world. Many of these new farmers would be going aboard the Arks to work on the farms to feed the thousands of people who worked and slept on the Arks as they sped through space. God had made the land fertile; woman and man had made the machines, tractors, diggers and combine harvesters to help the people feed themselves.

As promised by President Dmitriev, the army came in their thousands with farm workers from all over the world to prepare the land and get it ready for planting the food. The seed had appeared in giant containers, sealed until the land was ready for the seed to be planted. One of the new disciples was in each town, which now replaced the villages and the enormous camps were replaced with ten large towns, complete with shopping Malls, factories and houses, all fitted out with wooden floors, a cooker, microwave, electric kettle, fridge and washing machine in the kitchen. The bathroom was fully fitted with a shower and toilet, the bedrooms each had double and single beds, bedding and in every room in the house except the bathroom and kitchen, was a large wall screen which at the four prayer times throughout the day, 06:00, 09:00, 18:00 and 22:00, the televisions were two way and connected to the nearest church where if people could not get to the church for any specific reason, they could join in the main five minute prayer session and have their voices added to the millions of others as the whole world prayed to God.

If you were out, you could also join in the prayer session by calling the church, one number connected you to the nearest church and you could see and be seen by the disciple or vicar taking the service. Some people said it was not right to have to pray this number of times a day; the disciples did not even need to intervene as gangs of youths who heard what was said, intervened and explained in song what God had done for the people of Earth and what She was doing to send them into space was well worth 20 minutes a day. You only needed to attend two prayer sessions a day, but it helped if you could participate in all four. After hearing this, the people usually changed their minds.

Miraculously, babies and people who were in severe pain through an accident, were free from pain and the babies slept through every prayer session and the children between new born and five, were always quiet twenty minutes before and after the prayer session as the parents needed to prepare themselves or get to a church for the prayer times. On the Sabbath, it was as before, prayer sessions and a longer service where ministers and disciples led the service and told stories from the new bible to their flock. Once a month there was an update from the spaceyards, a direct live feed from the church to the spaceyards showing the people throughout the world how work was progressing on the Arks and sometimes, there were special tours of the Ark, the bridge, engine room, living quarters, restaurants and how people would live and work on the ships.

In Israel, Concord Aid had grown from the small building in the beginning, to a very large complex that was now covering half a square kilometre. The number of staff now totalled eight thousand and the complex was divided into five sections, Health, Farm Aid, Church Aid, Prayer and Space. The space section, although the people were not yet going into space, was always very busy. They covered the ISS, the Space Wheel, The Moon, Mars and Shuttle craft that were operating in local space.

Four new Sleuths, the Oyster, Eagle, Crane and Pelican, were being handed over to NASA, the RSA, (Russian Space Authority), ESA (European Space Agency) and the latest Space Agency EUROPA. (Earth United Operations for Peace Association). It was run by the church and the new Arks would come under that wing as well.

Each ship given to the various space agencies on Earth carried the name Europa and the church decided what to call the ships. If a really special ship was needed, then it could be named after the project it was being used for, but at the moment, old fashioned satellites, sent out on their long journeys lasting years just to get to Jupiter, were no longer needed. The four new ships were being used to travel as far as Pluto which using their new engines could get there in just under three hours.

The new Sleuths were quite large, 2000 feet long, 650 wide, and 400 feet high, carrying a crew of three hundred and eighty, with room for a lot of experiments and could carry one small sleuth craft, a squat ship, sixty feet long, ten feet in diameter, with its wings retracted for storage in the main ship, with the wings deployed, it was forty feet wide at its wing tips. The pilot could land the ship on an asteroid, get out, collect samples, store them and return to the ship. It could also carry a number of experiments to be deployed by the pilot who also carried a hand held anti-gravity projector so that heavy experiments could easily be moved from the ship to the place where they were needed to be set up.

The ship also carried a small shuttle able to carry ten crew and two pilots and a large number of experiments. It was also used to go into the upper atmosphere of Jupiter and Saturn to collect samples of gases. These were only interplanetary craft and would eventually be upgraded as the engines got better and faster. For now, a quick trip around the solar system, visiting every planet, could be done in two days.

There was also a call for a very large telescope to be placed further away from Earth where the sun was not so bright and the telescopes could see very far into space, looking back through time as the light travelled towards them. These were slowly being built and would be taken into space by one of the larger shuttles and deployed near Pluto with powerful transmitters to boost the signals back to Earth within a few seconds instead of hours.

Each time a ship was handed over, it was blessed by a disciple and there was something now happening every week as the people were getting used to their new jobs and the parts were being brought in faster to make the engines.

Everything about the new space engines was made as simple as possible and the wooden patterns for the engines components were designed by the women engineers, given to a disciple who was gifted with being able to alter the structure of molecules and in a few minutes the pattern was made and sent to the foundry where the parts would be made by their hundreds.

This speeded up the procedure and sometimes these disciples were called to the Arks to help change a design that was not fitting right. Once the design was changed, it was modified on the drawings and new patterns were made, the entire process taking less than an hour. In the main shipyards, women and men were now building the furniture to go inside the Arks, the beds, bunks, tables and chairs for the rooms, the tables and chairs for the dining rooms and the kitchen equipment to prepare the meals for the crew.

When the four new Sleuths were handed over, they were put to use almost immediately, the pilot was already trained to fly the ship and the captain and chief engineer knew their jobs backwards. Whenever a ship went into space, one person it would always carry was a disciple, and it did not mean the same disciple had to go on every trip; they could change ships and go when they liked. The disciples were part of the crew and had their own duties to carry out. The first place these ships were going for NASA, RSA and the ESA was to the Asteroid Belt where scientists wanted to explore the asteroids and see what metals they had which could assist them in making better ships and equipment.

The EUROPA ship Osprey, was heading to Mars, with a further 250 flat packed home modules. There were ten thousand of these flat pack modules ready to take to Mars and the Osprey was being used to transport the lot to Mars. The people there still had to wear space suits and were taking a long time to erect the homes, but now they came flat packed, it was a lot easier. When they constructed a further fifty homes, they were promised more help, but the homes would have to be built first; nobody had thought until now of giving the people on Mars the help that the ship builders had in the spaceyards.

When Osprey landed on Mars, two disciples were the first people to leave the ship. As the large hold doors opened, forklifts and trucks carried the first twenty homes to the place where they would be built. It was a nice day on Mars, the sky was bright red and it was a pleasant 10°C, you could see for miles as the wind was very low and there were not many dust storms in their area. The Martian people were amazed to see not one, but four forklifts and six trucks to transport the homes to their building plots.

Until now they had been doing all this with their small Martian buggies and one low level trailer which could only take one section of a house at a time. Now they were moving ten homes at a time and the trucks and forklifts would be staying with them. There were also another ten smaller electric buggies to help the people get around and a large number of experiments MARS ONE scientists had wanted to get to Mars, but as of yet nobody could get them there. Only four shuttle trips had been allowed to MARS ONE and they had been waiting for some time to get their own shuttle. Now their own much larger shuttle would be with them during the next month, much to the annoyance of NASA engineers who also wanted a large shuttle. NASA had for years been the forerunner of space research, but the Russians did not give as much Internet time to their followers, so were not seen to be as far advanced as they really were. Now they were showing their true colours in producing their large ships for interstellar travel, something NASA and the Americans were falling behind in.

As the work progressed on Mars, in India, work was almost completed on the two Terraforming machines. Another few days and they would be ready for their first test flight, a mere hundred metres off the ground.

Mary had been looking around the planet for where best to put her next project, the Trade Government for the species who would come to Earth and live here in peace and quiet. She wondered where she could put the complex and as a last resort called First Lady Rocc and Father Langley back home.

They were given an hour to meet with their families who they had not seen properly for three weeks, only occasionally being able to teleport or the new terminology, Jump, back to their rooms to spend a night of romance with their partners. When they entered the small lounge on level two, Mary was waiting for them with a hot pot of coffee and refreshments.

"Please, come in and take a seat," she said indicating the chairs around the small coffee table. She poured their drinks and took hers sipping the lovely hot liquid before speaking again.

"I have called you both here to discuss the next project; I want to build the complex to house the Interplanetary Trade Centre, but alas I am not sure where it should go or what country to put it in. The problem is the complex will need to be large; it will also have to accommodate a space port for their ships coming and going to Earth. We will also have to house a large number of their ships. The other thing I do not wish to do is block off a lot of air space as some of these people's ships will be very large indeed and need flight paths that will not take them into the current air traffic lanes.

"I am therefore asking you for ideas of where the Trade Delegation could go which will not offend anyone and also, stop people from getting too close to the buildings to see the new species. We don't want these people, some of whom are large lizards and use telepathy and walk upright, like the Leopards do, which is pronounced Leo-pards. These are animal people, they have the head in the general shape and colour of a leopard but have legs and arms of humans, but their feet are large and paw like, though they do wear shoes and have two arms, much stronger and longer than yours, with large hands and six digits for fingers. You will look just as different to them as they will look to you, but just remember they are all God's people," Mary explained.

Both disciples looked at each other in wonder, neither had any idea of where they could build such a place. In the corner of the room was a large, metre diameter world globe and another globe of Mars. Raymond, Father Langley's first name, looked it and with his mind, brought the globe of the Earth towards the table. Jean and Mary moved their cups and placed the tray of hot coffee onto a smaller table that Mary brought from beneath the main coffee table. Placing the globe on the coffee table, they watched as Raymond rotated the globe with his mind. There was plenty of room in the USA and again in North Africa, but when he concentrated and added the plane routes over the globe, they could see that it would be difficult for large spaceships to land in America where the airlines flew their planes every day of the week. It was the same for Canada and Europe, seeing the number of aircraft in the air at any one time; he was amazed there were not many more accidents.

Both poles were out of the question as it was now far too cold there, especially as God had mended the Ozone layer and alerted the seasons to how they should be. However, Australia was very large and only had a few airline routes going over the continent. Small planes travelled over the vast open areas of Australia, but they would not be bothered by large spaceships coming down from the skies and they would be well clear of the new buildings, which would require a runway as well.

"What about Australia, it's big, desolate in many areas, has a lot of open ground and you have plenty of room; I was thinking of Western Australia, somewhere near Lake Mackay. Could you or God alter the temperature around there and change the land?" Raymond asked.

"It would not be difficult, tell me more of your idea," Mary asked.

He took a sip of his coffee and then drained the cup thinking about his plan, getting it clear in his head as Mary refilled his cup.

"Well to put it in its simplest terms, Australia is a huge place, there are very few airlines there, the ships could come in over the South Pole and enter Western Australia from the Southern Ocean. Their flight path would be high enough so as not to interfere with public airlines and you would have enough room, if you could make the land around the area flat, to put as big as a spaceport as you needed, it could house a thousand ships and still not make a hole in the country. You would be able to have all your buildings joined together in a huge," he paused for a moment thinking how the complex would look.

"Ah yes, have it built in the shape of an octagon, so it could be seen from space. It could be as many levels high as you need and each wing or section would join to the next forming an octagon which would be at its widest, let's say for now two kilometres diameter. You would have all your staff work in one section, then as the people arrived, they would take up occupancy in different sections of the building and you would have in the centre a place for everyone to get together, a few flower gardens, some trees and the lakes would be used for recreation. If some of the people need to get into water, then the lakes would be there for them. There are also a lot of water sports that could be played there and everyone could join in and even have competitions if the other species can do this.

"The main thing about Australia and the position I have chosen, is that it will be well away from the public, it is hard to get to and there are no major roads and really, it's only accessible by air.

"We would have to build an airport and landing sites for the large spaceships, the ships could be placed in a hanger and we could have a normal runway for planes for visiting dignitaries from other countries and helicopters or space planes to take the visitors around the world. It might be an idea to build a few luxury space planes that have glass sections around the ship so the visitors can see our planet as they are ferried from one country to another," he added, quickly thinking of everything that had to be said.

"Oh one other thing, have you yet considered one government for this planet? I mean each county sends a representative to another place where the world is ruled jointly by one government, something like our current European Parliament. Except, this would be a world Parliament, where every country is represented and we have one government covering space travel and Interplanetary Trade agreements. In fact, it could be next door to the Trade Interplanetary Embassy," he said smiling to himself and thinking how the buildings would look and the space it would all take up.

"What do you think Jean?" Mary asked her.

Jean had been listening intently to what Raymond had said and was whole heartedly agreeing with him. The thing was; she could not come up with a better idea.

"I have to agree with Raymond, his idea is brilliant to say the least. If you join the world Parliament building in the same complex, interplanetary affairs can also be listened to at the same time and the trade delegations will represent their own planet. We too will have our own trade delegation and members from the world parliament will represent the Earth, it is a wonderful idea, an idea which will not only catch on, but will actually benefit the entire world."

"I am very pleased that Raymond has come up with the idea, I believe it will work and I have passed this idea and the plans as you Raymond, were telling me, to my Mother. A building this size will certainly do the job and will also need a trade delegation building on Mars as well as Parliamentary building at some time in the future, so I will ask her to take care of that at the same time.

"How do you see us creating a single parliament for the world?" Mary asked.

"That will not be difficult," Raymond said again, looking to Jean to get her approval before speaking. She nodded, as they most likely came up with the same idea, but Jean was quite happy for Raymond to give Mary his version first, she knew there were no favourite disciples here, everyone was treated the same.

"Each country around the world will still have its current leaders like every country in the European Market does. Each country will then elect a number of Trade Delegates and World Parliamentary Delegates; they will have free accommodation which we will need to build as well in some form of large hotel and be able to get them to and from their countries. It would also assist them if they could all speak multi languages and over time, create one planetary language something like American English which is currently used throughout the world. The Delegates for Trade and Interplanetary affairs will also need to speak the alien languages."

"That can easily be done, once again Raymond I think it is a very good idea and this can be carried out almost immediately. Do you agree with Raymond's ideas?" Mary asked Jean.

"Certainly, I was going to say the same thing, almost word for word," Jean replied.

"In that case, do your mind if I look into your minds to see your ideas and how they match up?" Mary asked.

They agreed immediately and Mary looked at their ideas, seeing they were almost identical and added it with her own ides as she understood them to be and passed the ideas to her Mother.

As they ate a proper meal and had another drink, Mary talked to them of other things she was planning and what was soon to happen when the ships were finally constructed.

"I have to be sure the Earth is at peace, there are still a few drug Barons who are against me and have tried their best to get opium and other drugs onto the market to make their money. As soon as they find the flowers to make the drugs, they disappear and so the workers give up and turn to God; but the men in charge are still not happy and will need to be convinced otherwise. This will have to be sorted before you both leave for the stars in six months time," Mary explained.

"Did I hear you right, in six months we leave for the stars?" Jean asked.

"Yes, the first ship will be ready in 13 weeks; after a week of space trials, it will be fully loaded with supplies. All being well, it will land back at the spaceyard at a different location being the main landing areas which I will prepare for them during the next month. The ships will have any alterations completed and then the crew will board with their families and they will get to know their way around the ship and be familiar with their own jobs. During this time their families will help to get the other cabins finished, cleaned and store the food and supplies to go into space. While the ship is on the apron, the outer paint work and names will be added to the ship.

"In the empty cradle, a new hull will be been given to the workers and they will commence working on it. To help the ship builders, I will give them the inner and outer hull of the ship together with all the superstructure and this time, if the first ship flies fine, I will help them in another way, but I am not saying what or how much will be done as of yet."

"This is excellent news, I didn't realise the work had progressed this far and so fast," Raymond sighed. He was looking forward to going into space and he was still not sure which ship he would be travelling on as nobody had been allocated a ship yet.

"Do not worry, I will be informing everyone of what ship they will travelling on and when they will be leaving as soon as the problems on Earth have all been solved, as of yet, we still have some work to do.

"What we don't need to worry about now is the Trade Delegation accommodation and the world Parliamentary complex, as this has been just been built," Mary said indicating the large wall screen behind them which had just come to life.

On it was an overhead view of Australia as seen from space, as the image increased in magnification they could see what was left of the Great Sandy Desert which had changed beyond all imagination. It was no longer a desert, but field upon field of lush green grass and trees which looked as if they had been there for hundreds of years. There were roads and paths leading out of the complex and two towns complete with shops and factories and houses for people to live in.

A huge eight lane motorway with automatic filling stations went from the new Trade complex to Perth, and another motorway to Alice Springs, which was now a huge international city with its own large airport, houses, Malls and shops. People who had once lived in North Africa and had been dying, now occupied the new houses, and were well, and could speak Australian like a native, were now working in the huge city. People had been taken from their homes from all over the world and intermixed with these people to show the Trade Delegates from other worlds that we could live in peace and get on together.

The two local towns would be the life line for the parliamentary and trade complexes. The people had been found to fill the homes and would be working at the new complexes, doing cleaning, cooking, helping to ferry to people around Australia and to other countries.

There was still a great amount of work to do and the people who now worked there, all seemed to know their jobs and what they had to do. There was not one single objection from those people who had been plucked from their homes and put down in Australia with all their papers approved and intact. The new immigrants had all been officially made Australian and now carried dual citizenship passports and everyone seemed very happy to be there working in a respectable job.

"How are you going to get the politicians to fill their places?" Jean asked Mary

"I have an idea about that and I will be seeing to this job personally, and I also need to put down the drug Barons. In the meantime, I would like it if you could oversee the last few thousand people in Brazil who have been hiding or kept prisoner underground by the drug

Lords and Barons there, it will be dangerous work, but I have faith in the two of you. You will both be fully protected so please do not fear death, the bullets, if their guns work, will not pass through your bodies, not only that, but you will be able to disarm them with your minds.

"There are eighty underground work areas holding over two thousand people, I will give you both their exact locations and I would like you to undertake this work immediately, it should take no longer than a week or two, maybe less of you can see the right people. I will deal with the governments while you take care of the drug workers. Can we agree on this?"

"Of course Mary, we always have faith in you and God," Jean replied.

"Then come and kneel before me for a moment so that I may bless you," she replied.

They put their cups down and the globe was returned to its former standing place. They both knelt side by side before Mary who stood over them. She placed her hands on their heads and said a prayer, both felt their body tingle from head to foot as they were given the special gifts; when she lifted her hands they both kissed her ring and stood up.

"I have given you my protection, and you will not be harmed by these people. You will find now that you know their locations and will be able to jump right into their underground places of work, which are hot, poorly lit and dangerous to be in. When you enter the room, say the prayer I have given you and the drugs will change to talcum powder. You will be able to put the workers under your control quite easily and make them understand what you are saying; the people who guard them will not be so easy to overcome. May peace and understand follow you in your time of great need," Mary said.

"Thank you Mary," Jean said and bowed her head as Raymond followed what she had done.

The two disciples returned to their rooms and said their farewells to their families once again before meeting in the small church where they both prayed for help and guidance before they left on their mission.

They both materialised thirty feet down under the ground in the hot Brazilian forests. Before anyone realised they were there, they had said their prayer and the drugs turned to harmless talcum powder. The guns the guards had were turned to chocolate which started to melt almost immediately as the guards held their guns firmly in their hot, sweaty hands in a threat to the people who worked there to get on with their jobs. The guards hadn't noticed the two strangers until they walked out of the shadows and spoke in their own language, which threw them even more, making them think they were people sent from their boss.

First Father Langley put both his hands in the air and everyone remained in their position unable to move. All they could do was listen to what he had to say while First Lady Rocc walked around the room and examined the tables and checked for other people who were further back inside the bunker. As she passed the guards, she picked a piece of chocolate from the gun and ate it, then pushed another piece into the mouth of the guard and allowed him to eat it but that was all he could do as First Father Langley gave his sermon and started to explain about God.

When his sermon was completed, he allowed the people to move, just a bit, and the guards found their hands were covered in the rich milk chocolate, they didn't know why, but it tasted so good they were passing it around to the people in the room eagerly, saying how good it tasted. When they all had their fill of the chocolate, they were much calmer and listened again to what First Lady Rocc had to say and how their lives would change if they would only just believe in the True God.

It was not hard work to get the workers to listen to them, they talked openly about similar pits like theirs, but had no idea of their locations, apparently the drug Lord for this area kept his people and guards knowing nothing about anything else he was up to or where the other sites were, including the drug Lord himself.

Three hours later they had the ninety people in the underground pit following God and believing in Her and willing to do as they said. The four guards opened the door and got everyone out, then First Lady Rocc said another prayer and the hole was filled with good earth and the tables and equipment inside, turned to dust.

All the men were from one village and allowing First Lady Rocc to read his mind, the group materialised a moment later five miles away in the middle of their village where they told their partners and children about God and what God had done for them.

It only took an hour for the whole village to agree to follow God and taking the village Priest aside, they made him a disciple and gifted him some powers so he knew how to pray to the new God and the times of day. First Father Langley promised them they would return later after they had visited a few more places.

One of the guards tried to contact his boss on the radio several times but they were not answering and as the guard explained, he had no idea what could be stopping them as they always picked up the radio on the second or third call. As First Lady Rocc looked to First Father Langley, they both had a fair idea of what the answer could be and smiled at each other exchanging similar thoughts.

It took a further two days to visit the other sites but as word was spreading, their job was made much easier and the last of the people joined the new church. As the two disciples returned through the villages, they moved the people together and with God's help, had a New Church built to allow them to pray in. They now worked together and were given a new road to the nearest city where they had the chance to start again. The village was now a small town with a Mall, doctor's surgery, chemist and other people to help them become decent citizens. If they wanted to move to the main city, they only had to ask Mary for help and help would be given.

Many of the trees were moved and the land opened up so they could use it to grow food and sell it in the next city to the people who needed the food there. Within a week the people had moved from being under guard and threat from the drug Lords, to having decent houses and jobs they could earn a good living from with the help of God. There was also a school for the children and a large telescope so the people could learn about space and the sciences which would be needed if they chose to travel on one of the large Arks in the future.

**

When the two disciples left Mary, she immediately put her own plan into action. Using her own knowledge of how the universe worked, and following her Mother's idea of creating deeper zones, Mary had created one of her own dark zones deep within the universe they were in. It had taken her five nights to create the zones within her Mother's universe where she had used everything she had learnt from her Mother and her own Master's Degree in Spacial Creation, to create this special zone where there was only one way in and one way out.

Leaving the Mansion of Disciples, Mary materialised deep inside the inner zones that were the pathways between the zones which separated the universe and galaxies. The entrance to her zone was well illuminated but as she moved deeper into it, the colours changed to blue, mauves, greens and oranges. As she moved further away from the entrance, the colours changed again to gold, yellow and red.

Taking a deep breath; Mary closed her eyes and blew a huge gust of wind from her mouth. The cavern she was in was immediately made larger and a second blow of breath made it larger again. Then she altered the colour of the cavern to a fiery red with a few whites and yellows which intermixed with the reds making it look like flames. There was one large window which looked out to nothing but blackness. The black was blacker than any known substance and when you looked into it for more than three seconds, your eyes saw an inner depth which went on and on for all eternity. You could then see a black object, falling end over end, falling into the nothing which filled the void outside the room. Mary then called upon the angels who were helping her at the Mansion, for they had offered their services after witnessing what some of the presidents and ministers had done to them and the threats they had made upon their lives and the lives of the disciples.

Somewhere else, the angels were warming up their voices for the next phase of Mary's plan. She wanted to bring the governments of the world, the drug Barons and drug Lords to a peaceful ending. They would either agree to join her and do exactly as she said or they would end up in the void on the other side of the window.

She was hoping to make them feel the fear their own people felt when going to war, make them feel how the soldiers felt as bullets flew over their heads and almost killed them. As the enemy moved towards them, they were safe in their houses, now they would be on the front line and know exactly how their own soldier's sailors and airmen felt when facing their enemy and death itself.

Mary closed her eyes when she was happy with what she had created within her own zone, not her Mother's zone, but hers. Closing her eyes, she concentrated, sending a message to her angels and then concentrated on the leaders of the world. Kings, Queens, Prime Ministers, leaders of the Opposition parties, Dictators, Presidents, Drug Barons, Drug Lords, and any Lord, Prince, Princess, Dame or Earl; anyone in the world's Monarchy who should be in the cavern with her; who had showed the slightest objection to Mary and her disciples. There would be others who had helped her, but would need to witness her power and agree to the new ways she was proposing or chose to change their minds.

Opening her hands wide, seeing each person in her mind's eye, Mary gathered them all together and brought them to the entrance of her *Death Zone*. At first it was pitch black, nobody could see anything or anyone else, but they could hear voices so they knew they were not alone. They could also understand each other and as they floated in nothing at the entrance to the *Death Zone*, where angelic voices started to hum very quietly, filling the air around them. There were just over two thousand people in the zone and everyone was just as frightened as the next person, although they would not like to admit it. Their fear started to escape from their bodies as sweat and everyone could smell the fear of the next person floating next to them. As they kicked their legs they wondered what was holding them upright and if they were in some form of bubble out in space and nobody knew how they got here; or where here was.

All everyone could hear were a lot of voices humming; over two thousand more voices joined the throng of low voices, creating the sound of thumping drums, bass drums, which echoed through the zone. As the sound of higher voices echoed around them, the zone before them started grow brighter with blues, hews of greens, mauves and orange and red. A loud voice erupted behind them, a high pitched scream that echoed around the *Death Zone* and through their bodies made them all jump, practically out of their skins. Without doing a thing, the world leaders were drawn further into the *Death Zone* as more voices, sounding like drums filled the zone. Four thousand angels cried out in agonising pain making everyone turn to look behind them as more voices added to the throng of beating drums, and thumping that was now filling the *Death Zone*.

The group of people were moving slowly through the *Death Zone* and they could see something large ahead of them, it was moving, coming towards them, flying through the air, almost translucent in the red glows that encompassed them. A moment later, everyone could see and hear the screams and cries of a millions soldiers who had been blown apart, with missing legs, arms, a headless image of a body passing over their heads and then dropping towards them. Their screams and cries of pain hammered around them, penetrating their bodies.

King William looked up seeing a soldier he saw die in Afghanistan come towards him. His legs had been blown off and his guts were hanging out of his body, with blood still dripping from his open wounds; when they were together there was nothing he could do but comfort him. Now here he was; passing right through his face, touching his soul like he touched the souls of everyone else who he passed through. He made them feel sick, as his penetrating cries of pain and agony filled their minds'. A sound of high pitched voices moving up and down through the scales with loud thumping voices, singing dum, dum, dum, echoed again through their minds, making them cry and shiver and throwing their thoughts into chaos.

As the people were drawn deeper into the *Death Zone*, strong red flames were flickering up the walls, the cavern was getting hotter and hotter, something Mary was not expecting, then a voice entered her head which was not her Mother's. Mary turned and moved in the *Death Zone* as the flames grew higher and higher up the walls like they were inside death itself.

Young children, six, seven and eight year old boys and girls, covered in a white powder, blood emerging from mouths, noses, eyes and ears, tears of blood running down their faces. A bullet hole in their head or stomach, allowed more deep red blood to emerge and drip down their dead bodies. These ghosts, with screams of high pitched voices and hums of banging getting louder and faster, rising and lowering in the octaves filled the drug Lords' bodies and hit their souls as the ghosts of those they had killed because they had seen too much or tried too much cocaine passed through them, tearing at their soul, trying to rip it from their bodies and take the soul with them to where death had no meaning and no time. It is an eternal night, and eternal darkness, waiting for the light of salvation to free them from their prison. They were now meeting those people that had put them where they did not want to go and all they wanted to do was exchange places with them, tear their soul from its body and put their soul into it, so they could live a new life and change the way of the people above ground for good.

As the flames erupted higher up the walls of the cavern, deeper voices added to the throng of angels and the two sides joined together, making an even louder noise filling the frightened visitors with dread and fear.

As the group of humans got closer to the cavern entrance, they could now see before them the red and black flames of Hell itself. As the first person entered the realms of the *Death Zone*, a booming voice filled the cavern. It was not a feminine voice but male, a deep voice which made their bodies shake with fear and dread. Some couples held hands, gripping each other for dear life, because now they knew what they were entering into. It was the home of the very person they had been worshiping all these years; they had been offered salvation but had refused to take it, pushing the offer back into the face of the person who had come to Earth to save them from the very God they were now about to meet.

They had welcomed this God into their fold, agreed He was much better than the new God who they feared more than the old God. Now they knew better, this God would take their souls and never give them back; there would no turning back now. The music and voices behind them was pushing them ever deeper into the red and black flames and smell of sulphur which burned when they breathed through their noses.

The angels behind them were now signing with high and very deep voices as the angels from Hell joined those from above. With the angels from Hell had come heavy large drums which were beaten with large sticks, men sweating with the heat and fear as they hit the drums, forcing the people to be drawn forward in the jaws of death where they knew they would be consumed by death and given to the Devil to do with as He pleased.

Their souls were now forfeit, they would never get them back, they could feel something within them trying very hard to escape through their mouths. The sulphur was making them feel sick so they would open their mouths allowing their soul to escape and try to hide or run back the way it had come and leave their bodies for the Grand Master of Death itself.

As the group were now into main entrance to the *Death Zone*, the heat from the walls of red and flames was overpowering, a thermostat showed the temperature to be 240°C and the temperature was getting hotter. The silver line was slowly moving up the gauge and according to the thermometer, there was plenty of room for it to go. With the loud drumming and combined voices of Heaven and Hell; those brought here had no option but move with the commands from the person who was pulling them closer and closer to Him. A large black silhouette was seen on the back wall through the flickering flames. A man with hoofs for feet, large hands, two horns protruding from his head and a full close cut, black beard around his face which was red and angry, an anger which had been growing for over a hundred million years.

He was now angry at what these people had been offered was thrown into the very face of someone He loved and cared for. When He saw her He felt love again, something He had forgotten over the millenniums. Now his heart was hurt and his anger was rising again. She had opened an old wound that would never heal.

Lucifer turned toward the people, opened his mouth and from it flew a million jet black ravens which flew through the group of people, flying right through their bodies and doing their best to shatter their soul and bring it to their master. Lucifer decided the best way to make these people listen to him was by a song as they liked music; so they would listen to his words in a song of rhyme. When Lucifer sung his first words, the combined angelic voices screamed an unearthly death cry.

"Come you masters of war," he commanded and the group of people shuddered because they knew he was speaking directly to them. Who else would Lucifer be speaking to and these people they knew were the masters of war.

"You build all the bombs; you build the big guns; you build the death planes; that kill all my sons. You hide behind walls; you hide behind desks; I just want you to know; I can see through your masks. You never do nothing; but build to destroy; you play with My world; like

it's your little toy," He shouted, His words penetrating their bodies, with the humming and harmonic voices getting louder and louder, filling the *Death Zone* they were now truly inside.

"You put a gun in my hand," He continued, making it sound like they had done this to Him. *"You hide from my eyes; you turn and run farther; when the fast bullets fly. Like a Judas of old; you lie and deceive; a world war can be won; you want ME to believe."*

A very loud scream travelled through everyone's body, making their very soul shake and the people knew for the first time they really did have a soul and it was not meant to end up down in the depths of Hell, it was supposed to return to Heaven from where it had come. It was their decisions that had turned their soul, and now they were feeling very sorry for themselves as they knew it was they who would never leave this place of horror.

"I see through your eyes; and I see through your brain; like I see through the water that runs down My drains. You fasten the triggers; for others to fire; and then you sit back and watch when the death count gets higher. You hide in your mansions; till the young people's blood; flows out of their bodies; and is buried in mud. You've thrown the worst fear; that can ever be hurled; fear to bring children, into My world," He sang at the top of His voice as the drumming and the sound of crying angels screamed in the Master's ears.

"For threatening my babies; unborn and unnamed; you're not worth the blood; that runs in your veins. How much do I know? To talk out of turn; you might say I'm young; you might say unlearned. There's one thing I know; though I'm older than you; that not even Jesus, would forgive you."

The noise around them was getting louder and louder, the voices and drums getting closer and closer. The hairs were now standing up on the back of their heads, for both men and women, were frightened to death.

"Death is much closer, than you ever knew, it is passing right through you, to a place it well knew. Death's waiting for you, just ahead of the queue, it's going to consume you, and everyone knew. As Death gets much closer, you'll see its huge Scythe, its black hood before you, covering Death's eyes. But as Death looks up; to the people He knows, you'll just stand staring, unable to move."

A huge scream came out of the angel's mouths, joining together making the people cry in fear. Lucifer stopped for a few beats as the angels cried out, humming and beating the drums all around the *Master's of War*.

"Let me ask you one question; is your money that good? Will it buy you forgiveness; do you think that it should? I think you will find; when death takes its toll; all the money you made; won't buy back your soul."

On hearing these words, everyone was clutching each other for fear was rising inside them, they could feel their very soul trying to escape from their body while it knew good angels were close by and there just might be the chance it could catch a ride with one of the good angels before Death consumed the body alone and let it rot for billions of years in Hell itself. The room was getting hotter and hotter, as they watched the flames reached the ceiling and moved around the walls, lapping the roof, getting closer and closer to the group of world leaders that had done so much wrong to the people of Earth. This was their fault; it was them who made the decisions and nobody else. It was not the man in the street that forced the country to go to war. It was those in charge who made the decisions to say we won't let that happen, we'll fight rather than talk and settle the argument over a meal and a table, taking as long as it would take to get the problem solved. It was not their lives on the chopping block, it was other people who they would never see and never know.

It was not actually them that fought the wars, Lucifer was correct in what He said, they ordered the guns to be built, found the money to pay for them and gave the money to people who profited from war which happened all the time. It was up to them to stop it, the drug Lords were the same, they made their money and they knew the consequences. They didn't even take the drugs, but forced the drugs onto others who couldn't pay the exorbitant prices and ended up stealing to get their fix and killed people in the process. Had they not bothered to make and sell the drugs, then people wouldn't be in the position they were before Mary arrived and healed everyone.

They were again pushed forward towards Lucifer with the angelic voices from above and below. They were getting closer to Lucifer who they did not wish to see or meet; but they had been praying to Him for two thousand years; they should have expected the worst to happen. They had just met Mary and told her to leave them alone, now they were about the meet her Father, who they wanted to see, except they didn't believe it would be here in Hell itself. They booted the good God off their planet and welcomed Her Husband into their lives with open arms; now they knew the truth, they wanted the truth to stop here and now.

A new figure came before them, tall, with a hooded cloak carrying a scythe in its right hand. Death lifted its outstretched hand and with a long wrinkled finger, with black finger nails, pointed to the window which turned so everyone, including those people at the back and in the middle could see through it into the black abyss outside.

As the window turned, a very loud scream from the angles filled the room making the people shudder. Everyone was forced to look into the black abyss, where they could see the depth of field behind it and the people falling forever through time into nothing. Now the group of world leaders where really frightened, fearing their own lives and their future as if they were standing at the front lines and seeing the explosions before them from enemy troops.

. Lucifer sung again with a booming voice that made everyone jump. His voice was now very loud and hard, threatening and although they didn't know it, feared He was getting close to the end of His song.

"I hope that you die; and your death will come soon; I'll follow your casket by the pale afternoon. I'll watch while you're lowered; down to your deathbed; I'll stand over your grave; till I'm sure that you're dead."

A large casket, with solid gold handles with the flag of Brazil draped over it started to float past the window. The lid slowly opened and Death turned to face one of the drug Lords. Death closed his hand and the person was taken through the air as the voices behind them changed from the drumming to an angelic cry of wow. The drug Lord was lifted, screaming and shouting through the air.

"No, please no, not me, it wasn't my fault, it was someone else, they wanted the drugs, they wanted them; I just supplied them. No please; don't kill me!" he continued to scream as his body levelled and passed through the glass window that separated the *Death Zone* from the abyss of death and time.

"You're mine!" Lucifer said and tore his soul from the man's body just as he passed through the window and his lifeless body was laid to rest inside the coffin. The coffin turned to face them and the lid closed. Lucifer held the man's soul, a small blue, sparkling sphere of pure energy, in his hand and threw it into the fires of Hell.

The coffin moved off, falling through time into a never ending abyss of blackness and death, it was now and forever inside the *Death Zone*. Another coffin sailed past the window and stopped. Cecília Ferreira, who was the second in command of one of the top guerrilla groups in the world, was chosen next. She turned her face and screamed at her leader Rodrigo Lopes.

"You told me to do those horrific things, you said everything would be fine and we should be prepared to die for our cause. This is not our cause and this is more than death itself; take my place . . . please, Rodrigo, you're the leader," Cecília screamed at the top of her voice as Death pulled her forward, her outstretched hands slipping from Rodrigo's clothes. He watched as she was taken from his side and turned his head away, only to have it turned back by Death to watch the grim death of his lover.

As Cecília Ferreira sailed towards the window, Lucifer reached out and took her soul, again throwing it into the flames of Hell. "You are mine!" Lucifer cried aloud as the angelic voices continued to praise Him and made the people in the group very frightened of what would happen to them.

Just as her soul was wrenched from her body, she looked back towards the group of people, her dead eyes still seeing their frightened faces. As she was placed into her coffin, her eyes beheld the lid closing upon her, she tried to scream in total fear but nothing would come forth, then as the lid closed, she felt her life force slipping from her body, she was dying and there was nowhere for her soul to go, it had already been taken by Lucifer and was at the moment on its long journey into the depths of Hell.

Death lifted his arm again and another man was picked, a Dictator this time, Honduras, Emmanuelle from the Republic of Honduras. He had ruled his people with an iron rod, killing thousands for pleasure, torturing them and sending them into battles to die for his glory to gain more land than he knew what to do with. He made his people steal for him, rob trains, forced them to put fear into others, even children.

He struggled to get free, holding with all his might onto a woman, who just happened to be Queen Kate. She pushed his hands away from her and gazed into his frightened eyes, knowing that very soon; she and her husband would be following them into the black abyss of death.

Again Lucifer grabbed the man's soul before he passed through the window into his own coffin, that was draped in his own flag which although would have looked good for his people on Earth, meant nothing to him now.

After two hundred and twenty three people were sent through the window, the angels continued to sing their songs, wail and cry as Death took their lives and Lucifer took their souls. Lucifer, with Death at His heels, moved forward into the red light of the cavern and faced the remainder of the frightened faces looking directly at Him.

"I have taken the lives of the worst offenders, it is now up to you to make a decision, can you all live in peace with your fellow people on the surface? If you cannot join together to form one united government, then I have plenty of room for you down here, just because you say yes now, does not mean I will not come and collect your soul when Death comes for you if you change your mind when you are back on Earth; as Death also serves My Wife, Death has never taken sides, it is Her job to do. Just so you know, you are not on Earth, and there is no way out of this *Death Zone*, which is thirty thousand light years from Earth, unless you are freed by the person who brought you here."

Once again everyone looked shocked at the word Lucifer had just said. Everyone had assumed incorrectly that Death was a man, now Lucifer was telling those assembled somewhere within the matrix of space and time, that Death was a woman. Nobody could understand this, why would it be a woman? There again, only ten months ago they were told God was female, they had assumed God was male but had not worked out that Lucifer was a God of equal importance and there were other God's who lived in another dimension where the souls came from. They still didn't know what purpose the souls had with God or where they were before they came to Earth. There was still so much to discover and so many questions to ask. Nobody knew how long they had to make their decision, nobody wanted to say a word. They were now being asked to form one world government but who would rule it? How would it work, nothing like this had ever been thought of before and so the question had never arisen?

"Lucifer," Queen Kate said and was pushed forward by calmer angelic voices to the very front of the crowd of leaders. Her husband, King William, tried to grab her arm but she was gone before he realised what had happened to her. He knew those who were at the front were more likely to be picked next by Death to meet their untimely end.

"What do you wish ask Kate?" He asked.

"We don't know how to sort this out, we have never had to have one government for the entire world, we have no idea how to even start the procedure to get it moving. Who would be in control or how we would make laws. The other problem is your daughter Mary is about to send our people out into space and there is talk that we will have extraterrestrial species come to Earth from other worlds to start trade agreements; we don't know what to expect with this either. We need help; can you help us or at least advise us on how to accomplish this and at the same time, save our souls?

"I am sure we all agree we have to move forward, but we need to be shown how to accomplish this by the right person. So what I am asking, as you have brought this up, would you at least talk to us without taking our souls or killing the rest of us to bring us to a peaceful conclusion?" she asked.

"I am not the person to ask for help, that person has been with you for the past ten months and you have not asked Mary for her help. I suggest now, you ask my daughter to give you the help and guidance you so desperately seek," Lucifer replied.

"If we agree to speak with Mary, will you release us from this place, wherever we are?"

"I do not have the authority or power to do this. For one thing, I do not know the way out. This is my daughter's domain not mine; it was her idea to bring you all here, but when I saw that She needed My help, I gave it like a father would help his daughter. However, I arrived here, wherever here is, from another route and I do not wish you to return with me until I am ready to take your souls. Should you follow me now, the dead in my company will grab your bodies and insert their souls in your place and you will be stuck forever with me inside Hell."

"This was Mary's idea?" Kate questioned.

"She needed to make you all see sense and also wanted you to live in peace, I suggest you ask her to help you and talk about this on the surface of your planet. Will you speak with her?"

"Yes of course we will," Kate replied eagerly, hoping this would end their stay in this *Death Zone*.

"Do you all agree to speak with Mary to bring this session to an end and live on My world in peace?"

"Yes," everyone said.

"I do not agree; I would like to be in control of the Earth Government, I think I could do a very good job and keep everyone in their place, so to speak," said Mr. Mason, one of the ministers from the American cabinet.

The President of the United States looked at him in astonishment and his spine shivered with fear that he would be called next to meet Death.

"It must be a universal decision, who else disagrees that talking to my daughter is not the right way forward?" Lucifer asked.

Nobody spoke and Mr. Mason spoke again. "You see, they do not have the guts to face you, they wish to take the easy way out. We must fight for what is right, not what you or God thinks for us; after all, it is our lives we are talking about. With me in charge, I will lead our people to make the right decisions and when we are threatened by these other species, we will exterminate them," he continued boldly.

"These are your own free thoughts and the way forward you would like to go?" Lucifer asked Mr. Mason; confused as to why he would say these things in light of what had happened in the previous four hours, not that time had any meaning here.

"It certainly is, Lucifer my friend; I am the main man and the way forward for the people of Earth. We do not need your daughter to protect us, we can do that ourselves and I'm sure your wife Mrs. God, or whatever her name is does not want you back in her life as you wish to be in control of the Earth. You keep Earth and we'll run it together, you and I and I'll take us forward and we can rule the universe with the ships Mary has given us," he replied smiling and looked around at everyone who just stared at him in disbelief.

"You idiot, we need Mary," Kate said scorning him.

"No we don't; Lucifer knows where he will get the best backing and I will be the one to turn the people of Earth and get them to follow Lucifer for the rest of time. Mary won't stand a chance against her Father. Lucifer has just admitted that He had to take control of this meeting over His daughter. I think Mary should return to her Mother's apron strings and stay wherever it is they are and I will rule the universe together with Lucifer. Isn't that right Lucifer?" he said grinning, full of enthusiasm for world domination and turned around to face the God he knew he would be able to trust.

Mary stepped forward from behind the wall of fire and smiled at Mr. Mason.

"I was sent here to help you, to bring you together and get you all to worship my Mother and allow you all to see the magical wonders She has made in the universe."

"I wasn't speaking to you, I was speaking to your Father and I see He has managed to trick you down here to give you the hiding of your life. You should grow up and do as your Father tells you," Mr. Mason said smugly.

Death moved forward and grabbed Mr. Mason by the arm and pulled him forward. "That's what I mean, bring me home Lucifer and we'll rule the universe together," he said; then his body shuddered as his soul was pulled out of him. He watched as the ball of bright blue light was cast down into Hell and turned to see the window approaching him and his coffin on the other side. The lid was open and then he was inside it with the lid remaining in the open position, the United States flag draped over his the rest of his coffin.

"I have decided to allow you to see your trip which will take eons, the long way around to meet up with your soul in Hell in about a hundred billion years. Don't worry, you will not need any substance to keep you alive, the *Death Zone* will do that for you; farewell Mr. Mason, your soul is already feeling the heat from my fires," Lucifer said.

"No, no," he cried as his coffin started on its long journey, falling through time into the cold, blackness of the *Death Zone*.

"Kate," Mary said in a loud controlling voice. "Are you still confident everyone will follow you and speak with me regarding the way forward and to form one government for the entire world to follow?"

Kate looked around at everyone, the angelic voices rose up an octave as Mary was taken higher over the bodies of the people behind her.

"Is there anyone else who wishes to speak about this single government with Mary? It is best you speak now, because once this project is set in motion, if anyone disagrees, I think Lucifer will be taking your soul without a moment's hesitation and we have just seen what Lucifer can do and where we could end up. I know there are still a few Dictators here, who have ruled their countries and people with an iron rod, but you too must join with us, you must also have a say in how the world is run and governed.

"I think before we move into space, we must be running this planet properly, together, a unified government making laws we all agree on. If we are squabbling among ourselves, just think how it will look to the species when they arrive here, they will think we are nothing more than children and should not be out in space until we have grown up. At the moment, if a new species came to our planet, we would all squabble who they should meet first and the country they visit will then be looked upon as if those people were given something extra that the other countries did not get.

"Now we moving into space, we must act like proper adults, we need to have one place for the people to meet us where each country is represented and we all know where we stand and what is on offer to us all, not just one country. I cannot speak for everyone here, but I can speak for myself, if it were up to me, I would be asking our government to accept the proposal to speak to Mary about this project and let Her give us Her ideas and wisdom. We should not forget that Mary is the daughter of God and Her Father is also a God; just because He is not with His wife, does not mean He does not care for His daughter and want only good for Her, so I think what Lucifer has done for Mary is no more than a father taking a hand and saying let me show you how it could be done, then go off and finish the job yourself," Kate explained then turned to face Mary and Lucifer, with her heart in her mouth, hoping she had said the right thing and not sent her soul into Hell.

"I think this also shows us, as leaders of the Earth, that Lucifer is also a very caring parent, He loves His daughter and does not want any harm to come to her. That although Mary conceived and built this *Death Zone*, possibly following in Her Mother's footsteps, Her Father was standing behind Her, holding Her hand in a metaphoric way saying to her I still love you despite what has happened between your parents, I am still here for you. Am I just a little right Lucifer?" Kate asked, now fearing she might have gone a little too far and risked her own soul to go into damnation.

Mary turned to face her Father and their eyes made contact; silence followed for a very long minute, or at least it appeared to. What was said between them only they were privileged to know.

"Kate, you are right in everything you say, I still have feelings for my daughter and my son Jesus, Mark has remained with me after I took him when we split up and went our own ways many eons ago. When I first saw Mary arrive at Radstock Church, I needed to see her and touch her and with the aid and permission, which Mike Adams has kept secret all this time, Mike allowed me to use his body to see my daughter through his eyes and I used his hand to touch her skin, which to me at that time, made me feel so unbelievably whole again. There had been something missing from my life and now I know what it is.

"You are right, I was protecting my daughter, not that she needed it, she was quite capable of making you all see sense, but she was not able to discard the souls which I have taken. Neither would the voices of angels been so dramatic had not my angels joined with those from my Wife. A few of you needed to go and I had my eye on them for a long time, they are now where they need to be.

"If Mary agrees to your request for help, and if she releases you from this *Death Zone*, I will need to keep her here for a short time to speak with her about my own feelings and a time to catch up, if you know what I mean. As far as I am concerned, I am now releasing you all into my daughter's care and it is now up to her as to what she does."

Kate looked to Mary, hoping they would not be kept here any longer than needed.

"Kate, I am very pleased to hear what you have said and you said it with such conviction and hope for your people. For the discussions that will follow, I will talk with you all in better surroundings at my Mansion of Disciples. For the benefit of the leaders here, I will be in charge of the meeting and give you my ideas of a way forward, you Kate, will be the spokesperson in the talks, for you are the only person to speak up and you have told everyone here the reasons why you should move forward and have bowed to a higher authority and asked for help and guidance. I will now release you all from this place and send you to my Mansion of Disciples.

"You will arrive in one of the great halls where you will find my people in attendance. There will be food and drinks available to you but no alcohol, there will be no large table as I do not think this will be necessary and you may take whatever seat you like. At the back or front, you will all be able to be seen and heard by me.

"Before you all leave, I want you to take a last good look around this place, look out of the window, have no fear, you will not be passing through it today, tomorrow, who knows what the future will be? Once you have all had a chance to look around for one last time and looked through the window, you will be allowed to leave: you will appear at my Mansion of Disciples, but you will not leave until you have looked through this window. Now, if you will excuse me, I need to speak with my Father, but I will not be far from you; have no fear of that," Mary said.

Everyone was then able to move as the angelic voices hummed in the background, now giving a relaxing composition of voices and music to try and soothe the people still alive in the *Death Zone*.

William pushed through the crowd and took hold of Kate's hands, holding her close and kissing her before speaking to her and walking with her, hand in hand to the window. As they peered into the black *Death Zone*, they could still see Mr. Mason, now sat up in his coffin crying with fear, mumbling to himself as he fell into nothing, falling for eons into nothing until he finally arrived with Lucifer and joined with his soul that was at this moment, waiting for him, angry that it had been cast out and treated in such a dismal way. The soul should now be on its return journey to Heaven, not waiting more eons for that to happen.

As Kate and William looked through the window they knew what could be waiting for them if they did not agree to, or at least listen to Mary's ideas for their future. Turning around, they looked at the flames flickering up the walls and over the ceiling and listened to the angelic voices. The next moment they were gone from the *Death Zone* and arrived in the bright light of Mary's Mansion of Disciples. It was a very large room and as they set foot on the floor, people were approaching them with a choice of hot drinks or pure, cold water. There was a large table full of hot and cold food for them to eat and within a few minutes more people had joined them, obviously seeing everything that Mary had told them to see.

Mary was now in a cooler room, sat down facing her father who was holding her hands in his. His face had changed to one she knew better, there were no horns coming from his head; it was all there for the people of Earth as that is how they imagined him to be. He was handsome, not showing his age of eighty nine thousand plus Utopian years at all; he looked fit and dressed in the usual clothes a male God would wear; a suit of the finest cloth, a coloured shirt, tie and blond hair. His electric blue eyes were the same colour as hers, and the angry red eyes for the appearance of the Earth people were now gone. He was normal again, to her anyway, and once again the father she had known when he left for a specific purpose which she was not privileged to know.

The secret had remained between her mother and her father, why Mark had left with him she had no idea, but she recalled begging him to stay with her. However, Mary was younger then, and much time had passed, or at least it had appeared so. She wondered now if her father would tell her why he was here, or would he keep the secret a while longer? She had the feeling he wanted to come home, at least she hoped he did so they could be a family once again.

"I'm so pleased to see you again, to touch your hands, see you close up, talk to you, I wish the past had never happened, but I had a job to do and your Mother and I knew it had to be done if we wanted the answers to something very important."

"Why did you do it?"

"Do what?" he replied, unsure of what his daughter was talking about.

"Mess with Mother's pet project, this universe. Mother told me all about it, she said you interfered with her project, took the Earth and this Cosmos away from her and made it your own." She looked at her father as if she was speaking nonsense and had no idea what she was talking about.

"That is why you split up isn't it, because you stole Mother's ideas?" she asked, now unsure if this is what really happened.

"Ah, this is what your Mother told you?" he asked.

""Yes it is, there is so much confusion at the moment, the people are getting very concerned something bad is going to happen, Father, I'm sure Mother needs you back with her if what I have heard is really true."

"What, exactly have you heard?"

"Talk, people talk behind our backs, the Elders and scientists speak in riddles and when any of the children are close, they change the subject. We all know something is not right, but as usual, until our twenty thousandth year has passed, we are not allowed to know what is truly happening in our worlds."

"And that is right; we do not wish you to concern yourselves with adult affairs, when you have so much to learn. Listen to me, I was not jealous of your Mother; I failed universe design twice and the third time I just scrapped through. I was good at controlling matter, good at a lot of things, I was very good at making you and your brothers, or rather we both were, as it takes two make a child. I still love your Mother very much, but we no longer communicate and it is not easy with me down here to get a message through to your Mother to try and talk again."

"I could take a message to her if you really want to start speaking again."

"Yes, I would like that," Lucifer replied.

A thought came into Mary's head which she obeyed a moment later.

"I have to go; Mother wants to speak with you. It was lovely to see you again Father," she said and leaned forward to kiss his cheeks and lips. At the same time she gave his hands a big squeeze and sent a few thousand messages into his mind. The last one was very simple. "Tell her you still love her and ask for her forgiveness and try to get back together."

"Mark has been saying the same thing," he replied then Mary was gone, joining the world leaders at her Mansion of Disciples.

"So, you want to talk after all these years," Godisious said her husband.

He was stunned and shocked that his wife had appeared before him, this was the first time they had met since they parted. He stood, looked at her and sent several thoughts into her head like he used to do. They sat down together and Godisious brightened the room up and prepared drinks of nectar, their favourite drink which they consumed over the next twenty hours which flew by in seconds.

At the end of their talks, Godisious knew he was almost ready to return to her side, with what she had learnt from her children, she knew she too had to change her ways a little and meet her husband half way. She was about to go when Lucifer had an idea and called Mark from the depths of Hell. He was still unwell from his Mother's hard shove but he did as his Father had said and looking ill, and heavily bandaged around his torso, he appeared by his Father's side. When he beheld his Mother, he couldn't stop himself from walking forward and falling into her open arms. They kissed and embraced, like a son would after being parted for so long from his mother.

She looked him, and put her hand to her face upon seeing what she had done to him and apologised for his condition.

"Open your shirt," she said.

He did as his Mother said and she placed her hands over his torso and with all her might healed him, something his Father could not do as it had been done to him another God and only the God who had given you the punishment could take it away before the body very slowly healed itself, which could several hundred years, a very short time in a God's life.

Feeling better, with his broken bones healed, he moved his body around and got used to moving without pain again. He sat down on the seat his Mother provided and also drank the nectar.

"You're a bit young to drink this, even at nineteen thousand years old, but as it's a special occasion, I'll allow it this once, but not too much. She poured a small quantity of the holy nectar into his glass which appeared on the table before her and offered him the glass. "Do not drink it too fast," she warned him. "It is very potent."

Again they talked for a very long time, and when she was about to leave, Lucifer sent his wife a message using telepathy so their son didn't hear.

"Are you certain?" she replied speaking the words. She looked much happier almost jubilant.

"Yes, I am, it will not be long now I can assure you. Just a little more time is all it will take," he replied and Lucifer too looked happy, he was smiling now, something he had not done in front of Mark for some time. He had always looked concerned, as if something was not working right and their very lives were in jeopardy.

"Then let us hope what you started will be concluded very soon," Godisious replied.

"To show I am sincere in my thoughts, Mark, if you so desire, you may return to Utopia with your Mother, I will leave a doorway open for you to visit me if you so desire, but your Mother needs to see you and you are of an age when you should decide for yourself who it is you wish to be with."

Mark looked at his Father, not really believing what he had just heard. He had wanted to be with his Mother for many thousands years, but his Father had always denied his access to the upper realms because he needed help where he was to move the people of Earth forward. As soon as Mary appeared, he was a changed God and Mark knew his Father was sincere in what he said, as a God could not lie; it was not in their character.

"If you will allow me to still see my father, then I will join you at home, but Mother, I implore you to think again about Father, he has changed, and just now he helped Mary sort out the people of Earth."

"Oh yes, what about Mr. Mason, drifting through the *Dead Zone* for next few millennium?" She asked.

"He is dead yes, but that was all for show to make the people of Earth listen to our daughter, which is what I thought you would have wanted anyway."

"Yes I did, that was very thoughtful of you but there is still Mr. Mason."

"He is currently with me in Hell and laughing and joking with the other people I took from the group and then he met more of his friends who are helping me in my . . . experiments," he said not wishing to give too much away as Mark could hear every word they said.

"When you return, what are you going to do with the others who are assisting you and what of the few souls who you have down below?" she asked.

"I'll release them from their jobs right now, but there are many people with me so I will send them home a few thousand at time so as not to overload the welcoming crew," Lucifer suggested.

"Mark, what are you going to do?" his Mother asked.

"I will go with you, if you allow me to still see my Father. When we were young we were together, but for the last, I don't know how long, we have been apart, it was always so nice to have you both as our parents. What I'm trying to say is, when I was on Earth, Jesus spoke to me and asked me if I would ask Father to make amends and return to you which I have done. Mary asked the same of me just before you pushed me into the depths of Hell, which I well deserved for what I had done to ruin her day."

"I will still allow you to see your Father, now, do not worry about your parents, we will work things out I assure you, but it will take a little more time, as your father still has work to do here. It was never about us as to why your Father came here, He had a job to do and has done it well with those people who joined him. Believe it or not, and you must not breathe a word of what I tell you now to anyone at home, but your Father is doing His very best to save all our lives. If He had not come here, we would not be getting the answers we need so desperately to know.

"I know this sounds strange, but your Father is a very brave man and is doing what he can to help everyone on Utopia Prime and in our universe. Now, go join your brother Jesus, I'm sure you will see things have changed and when you're there, it will all become clear. I will meet you there when I have finished speaking with your Father," Godisious said truthfully. She kissed her son's cheek again, held his hand and sent him on his way.

"So what are we to do now?" Lucifer asked.

"Will you hold me like you used to?" she asked.

They embarrassed and kissed, long and passionately, passing thoughts between them as they rekindled their ancient romance. When they finally parted, they both felt different people, as something inside them had been healed.

"You may send all the souls home; I will make arrangements for their loved ones to meet with them. Close up Hell and join me when you have completed your work. How long will it take for you to send the souls to me and clear up the *Hell Zone*?"

"Would you like to keep it as it is or shall I change it to add another zone to the universe?"

"I would remove the heat and make it safer for those who discover it, and perhaps add a few feminine touches."

"I will change it and when I join you, perhaps we should allow Mary to change it, let her see what she can do and it will tell us if she has managed to learn anything from you, what do you think?"

"I think it's a good idea," She replied.

"Then I'll see you soon, I will send the souls over the next six days, so as not to overload the systems in place. I will still need some of my angels to assist me with the work I have to finish, and I cannot leave until I know for sure the project is fully underway," Lucifer replied.

"I understand, I know how important this project is, but you must admit, without my help and Mary's intervention, it may have gone the other way," Godisious replied, speaking up for their daughter.

"I know, I will tell her how much I appreciate her work when we are all together." They kissed each other again, embracing for the last time in the *Death Zone*.

Godisious returned home to make the preparations for the returning souls and Lucifer returned to Hell, where he immediately turned off the extra boilers and brought bright light to the zone. He called his angels and told them what was about to happen who rejoiced with songs of praise; knowing that very soon they would be joining their families once again.

Within four hours, the first twenty five thousand souls returned home, then more and more souls followed in their wake. A thousand of Lucifer's angels helped to clear up Hell and remove all the dangerous objects that were there to impress the souls coming down from above as they entered Hell's domain.

Every gambling casino was closed and cleared away; every house of ill repute was cleaned and given a face lift so his wife or daughter would not have to see what the humans had got up to in their spare time. The souls returned home in a first come first to go position and every soul wanted to meet their loved ones again as soon as possible.

As this was happening, Mary had her own job to accomplish.

CHAPTER 21

When Mary appeared in the room, the group of world leaders had been talking over the problem for several hours, which had also given them time to eat and drink and refresh themselves. They all had a chance to look around the lower rooms in the Mansion of Disciples and see some of the grounds.

"I'm pleased to see you are all talking and getting on, have you had a chance to look around?" She knew the answer, but thought it polite.

"Yes, thank you Mary, it is a very nice place you have here, I can't see what Mr. Templeton and the councils were objecting to," Kate replied.

"Neither did I, have you discussed what is to happen yet?" Mary asked Kate.

"I, like everyone else here has dropped their titles to stop making us look and feel too obstinacious. We have all eaten and are ready to start talking and listening to what advice you can give us on how to move forward. We have started to discuss the problem, but as of yet, we have not come up with a place where government could be placed," Kate replied.

"Before I answer that question, what did you think when you joined me in the *Death Zone*? You may as well get it over with before we start the talks, if anyone has anything to say, get it off your chest now, so we can leave that behind us."

"I have been told to say, that we didn't like the way we were treated, but we can see why it was done, we are the leaders of the world and we should be treated as such, but for that to happen, we should act like adults. Not only that, if we want to lead the country, we need to be able to represent the people and we were not doing that, we needed the kick up our backsides and you have managed to do that and make us see sense. We are now ready to listen to you, with our ears wide open and ready to do as you suggest," Kate replied.

"Thank you for your honesty, it is good to hear you didn't like the way you were treated but it had to be done. Now, without any further talk on that subject, let's get down to business," Mary replied and got everyone to take a seat.

Their talks lasted for twelve hours, they were then shown to their rooms and the discussions for the new government finalised the following day. When the leaders were ready and knew what they were going to do, everyone agreed they should give a press conference at Buckingham Palace. The press were called with Mike Adams taking the lead and asked the main questions of Queen Kate.

"Welcome everyone to the United Kingdom and Bucking Palace," Kate said when the time came for the cameras to roll.

"I am Queen Kate and I will be explaining what will happen and how we are going to form a single World Government. With Mary advising us, we, the leaders of the world, have come to an agreement that when our people travel into space and start meeting other species from other planets, we will be able to open new trade agreements with the species and for this to happen, we will be sending people from Earth to represent us and we will in turn have their people come to our planet to represent their species and start trade agreements. This means we will be trading between planets; which could be in the form of knowledge, health or technology; it could even be in rare gems or metals. In turn, we will offer them our technology, gems and ores from Earth and in our solar system.

"When these species come to our planet, we do not wish to start an interplanetary war just because they look nothing like us, but at the same time, we want to impress them and show the species what we can do and show them how we can accept them in our world. To accomplish this, we need to form one world government which will represent every country in the world. What we don't want to do is fight over what country the visitors will visit first and get the feeling that one country is getting more than the next, so by having one world government, we will all know what everyone is being offered and the trade will be spread equally around the world and every country will get the benefits of this trade."

"Your Majesty, if I may ask, that is all well and good, but how will you form this government and will we lose independent government of our own countries?" Mike asked.

"I'm glad you have brought this up, the one thing that will stop is the European Government. We have talked long and hard with Mary who has shown us which way to go. Like the European Government, we will still rule our own countries; each country will have a President, King or Queen. For the United Kingdom, we will be back in power and we will like before create a government to oversee our decrees. There will be so much less work for us to do with no one breaking the laws; the police will have a new role, helping people and controlling traffic. Very soon, all the old petrol and diesel cars and lorries will be gone, being replaced by clean electric vehicles.

"Our forces will still be needed except they will now help throughout the world to get this planet working properly, like the Russians have done with their army, navy and air force. Some of the troops will also be used to help make the very large spaceships and some of them will be going into deep space. A lot of the air force pilots will be used to fly the new space shuttles and smaller interceptor craft and inquisitor craft. These are small, very fast ships that will be used in space to take pictures of God's universe and explore asteroids, planets and even comets. They will also be used in accidents in space to help get people back on board ship if they fall off the ship while working outside while the ship is in flight. Does that answer your question Mike?" Kate asked.

"It answers some of it, but you still haven't explained how you intend to form the universal government," he replied.

"Well, we will rule the United Kingdom and the same leaders as before will rule their own countries, but the people of each country will choose who will represent this and every other country in the world. Once these people have been picked, they will work at the new United Parliament Complex."

"Where will that be and how will it be paid for?" Mike asked.

"Mary has put this question to her disciples and they have up with the answer. We will be using part of Australia; in particular, Western Australia. Mary has spoken with her Mother and God has already constructed the new buildings and the Interplanetary Trade Complex. This is near the Great Sandy Desert, which is no longer a desert. There are I believe eight lane motorways joining the two complexes to Perth and Alice Springs, which have both been made much larger. There is now a very big spaceport there and a huge complex to cover the incoming spaceships and repair or check them over for faults.

"Every building has already been made, furnished and people are now working in the two complexes getting them ready for when they will be needed. A large airport has also been made and so too have two large towns between Perth and the two new complexes, these are called Hope and Peace."

"Can I ask why Australia was picked and not America, after all, this is a country of the United Kingdom and people may think we had a hand in this decision and have persuaded Mary to favour us," Mike suggested.

"That is not so. It was chosen because of the large spaceships which will come to us; they will come in from orbit over the South Pole and land in Australia. Had they come down in Europe, Africa or America, there were too many aircraft lanes they would have to negotiate. So by coming in over the Antarctic Ice flows, they will not interfere with any aircraft and there are only a few planes a day flying into and out of Perth. The smaller aircraft which fly low over Australia will not interfere at all with the large spaceships and shuttles."

"How will we and the other countries pick our representatives?" Mike asked.

"We also put that question to Mary. What has been decided, that when this programme concludes, each country throughout the world will have the chance to select their own people to represent them. For this country, and it will be similar in every other country, the first ten people will be represented from the government. What will happen is the government ministers, from all parties, will stand on a stage before the cameras and tell their county why they should be picked. The people will then vote by thinking which candidate is best and Mary will count them. Everyone will have the chance to vote, even children over the age of twelve.

"When the ten government ministers have been selected, which will not take very long, people throughout each country will have the chance to put themselves into the people's parliament. All they have to do is think to Mary why they should be chosen; only twenty people per country will be selected and Mary will select the people, her decision will be final. They will then join the ten ministers on the stage and they will be told when they will take up their positions and what their duties will be. Those of you, who are chosen, will have the opportunity to take your families with you to Australia."

"The Trade Delegates will be selected the same way, only this time there will be many more people involved. The main Earth Trade Delegation will consist of twenty people from large companies who work and not just sit on committees; they will have an active part in their company and be selected by Mary.

"Then two hundred people from each country will be selected to be Trade Delegates who will travel far into space and remain on distant planets to represent the Earth when we form trade agreements with other species. For this, you must be dedicated and not mind remaining out in space for long periods of time and staying on a different planet. You will of course be allowed to take your families with you. You will not have to stay away from Earth all the time, but will be brought back here every four years by a fast Schooner of the latest design.

"When this talk is completed, you will be able to press the red button on your television handsets and go to the page where you will see all the spaceships available for use to the Trade Delegates and also, read what your duties will be. If you are considering seriously about being a delegate, then think to Mary first, she will then allow you to speed read through the entire brochure. Be warned, you will not get this gift if you are not sincere, and Mary will know if you are.

"If you are picked for any of these posts, Mary will gift you with speed reading, and multilanguage speech and understanding, some people will also be given the gift of telepathy, but not everyone. The reason we have decided to do it this way is to save time, as it will take months to sort this out in our previous ways of voting and we do not have that long, also, it will not cost us anything to do." Kate explained slowly and confidently.

"When do you expect the first of the space Arks to leave Earth?" Mike asked.

"Mary will be the best person to answer that question, but I think, the first ship will leave within four or five months. When the first ship leaves Earth, we must be ready to accept and start working on the trade negotiations," Kate replied.

"Are you happy with the way these votes will take place your Majesty?" Mike asked.

"Yes I am, very happy and confident it will go well. Now, I have just been told we are ready to start picking the UK's ten ministers who will represent us at the World's Parliament Complex. The reason we have chosen the UK to be first is that we are talking to you from Bucking Palace and we wanted to show the world how it would progress. Once this country has picked its ministers, the rest of the world will pick their ministers. I will now hand you over to Mr. Peter Stone who will tell you the names of our ministers up for selection," Kate explained.

The scene changed and within the hour, the new World Ministerial Government had been picked. An hour later, each country had ten civilian leaders to ensure the government ministers did not make any mistakes and did what their country expected of them. The Trade delegates were also picked the same way and after three long hours, everyone was chosen for the new posts.

The people were then taken by Mary and her disciples to see the new buildings they would be working in and after a four hour tour, were sat down and told what would be expected of them. The people would not take up their new posts for a further two months, and during this time, they would meet in their respective countries and get to know each other and what duties they would have to do.

Mary was now going over what her disciples had previously done and speaking to them all the time, reinforcing her faith in them and telling them where they could go and who to talk to. The prayer sessions were going extremely well and the best news she had, was that her Mother and Father were going to get back together soon and were now working to close Hell down. All the souls where now back in Heaven and reacquainting themselves with their departed relatives and friends.

The spaceships were now getting closer to completion, much faster than Mary had hoped. Three of the first four ships were at the same position, loading supplies and the last items aboard the ships. The fourth ship was only a week behind the other three and the people working on this ship in Salisbury Plain; were working as fast as possible to catch up. The fifth ship, known as Explorer, was also much further advanced than was expected at only a year into its build.

A further twenty of these six kilometre long ships were now being built throughout the world and thirty small Arks, four kilometres long were in the process of being constructed. There were now, a hundred spaceports throughout the world building much smaller ships that would be used to explore space and take scientists, astronomers and people around the cosmos to see God's wonders.

Each ship carried the very latest cameras and telescopes so the people could see even more of the universe. Anyone could put their name forward to go on the ships and if you had a trade which would help aboard the ship, you were able to push yourself higher up the lists to get on one of the ships. People now wanted to learn more about astronomy, astrometrics, engine propulsion and spaceship design in general. The medical profession were also needed aboard the ships and so too were cooks and chefs as people still needed to be fed.

Other spaceyards were concentrating on building the smaller Interplanetary ships which would explore the planets within our solar system and a few larger ships would travel out to Alpha Centauri and further. Six spaceyards were building the small one person ships and space shuttles which were used between the Moon and Mars.

There were now over forty shuttles in service already and a month earlier, the two very large Terraforming machines were launched from their spaceyards and sent to Mars. It took them a day to get to Mars and a day to get to the correct locations and land. Once they landed and were in potion, the two machines were manned, synchronised and started. Their job was to change the Martian poisonous atmosphere, to one the people of Earth could breathe so they would no longer have to wear their space suits and be able to walk around Mars like they did on Earth.

The machines could be heard all over Mars making their thumping and pulsating sounds. It would be a slow process and it could not be rushed, but they hoped within four weeks the new Martians would be at least be able to walk about for a few hours a day without their spacesuits. To terraform the entire atmosphere would take four to six months, but it would be completed giving the people of Earth a second planet they could call home.

The shuttles continued to ferry parts to Mars and the Osprey dropped off six large loads of homes bringing the total to twenty thousand homes on Mars. A small nuclear power station and two hundred electric cars, lorries and mechanical equipment was also sent to Mars to help get the houses built from their flat packs to liveable accommodation. On the Moon, there were over a hundred buildings and four companies had commenced drilling there, while scientists carried out experiments and put ten large telescopes on the Moon which had no atmosphere so the astronomers got brilliant views of space, day or night.

As Lucifer and Godisious grew closer together, the Earth's skies continued to get brighter and bluer, the cosmos became cleaner, with all the cosmic dust being filtered from the spaces between planets.

The four large space Arks lifted off within a day of each other and commenced their space trials. As soon as they were in orbit, God gave them a further fifty Space Arks that were only a few weeks from being completed and ready to leave Earth; these were parked on new aprons in ten different countries. In place of the Space Arks that were now starting their week long space trials, four new triple skinned hulls, complete with engines and generators were put into their empty cradles. Another fifty spaceyards were added around the Earth each with triple skinned hulls with engines and generators already in place. This meant the Earth would soon have a hundred and eight large Space Arks available to them.

God had also given the people a further forty very large World Ships. These were eight kilometres long, five kilometres wide and high. These would take the people from Earth and move them to new worlds which the first Space Arks would find, and start humans moving out into the cosmos, spreading their wings. Each ship was capable of making homes, factories and any other buildings from raw materials on a new planet. They also carried all the equipment they would need to start creating a decent world for the people to live on and expand.

This is where the current world forces would pull their weight, they would be the first people to go to the planets and check them out for dangerous wildlife and diseases. They would be the people to erect the first fortifications on the planet for the farmers and builders who would arrive a little later and get the planets ready for the settlers to come and live on the planets, moving the people from Earth far out into space.

The Terraforming machines would also be used on planets where the land was good, but the air needed cleaning up. The World Ships were big enough to take two Terraformers with them and drop them on a planet, leaving them to do their job once they had been set up and started. Six months later, the people would return to the planet and the Terraforming machines would be moved to the next planet the surveyors had found.

It was now that God needed her children to go far in space and see her wonders and spread the people around the cosmos, and for that to happen at speed, She would need to give them some help. Although She still gave the people jobs to do in building some of the ships, the others would get the people into space and started on their next phase to bring the huge experiment that Lucifer had started to a conclusion.

When the four space Arks, Enterprize, Dauntulous, Excalibur and Valiant landed, the fifth Ark Explorer, which would be going to the Andromeda galaxy, lifted off to start its own space trials. This time the Ark was followed by a large number of small ships and two large shuttles as the people wanted to see this enormous ship get off the ground and safely into orbit. This was the biggest ship they had ever built and everyone wanted to if it would fly.

When the giant ship lifted off, it got to a thousand feet and then the rear end of the ship dropped down and the nose was pointing at a thirty degree angle to the sky; there was a huge cry of "NO!" from all over the world as everyone thought the massive ship would crash through lack of power.

As the helm officer, Miss. Jenifer Jackson, compensated for the extra bow lift, she turned to face Captain Veronica Scott from the UK, blushing profusely. "Sorry about that Mam," she said, "Holding position."

"Is that what you call it? This is very embarrassing, shall we go? That way if possible," she said and smiling and pointed to the sky which they could see on the large view screen before them, which went right around the forward bridge curved bulkhead.

Below them, on the lower bridge, was a clear glass window four hundred metres long and ten metres high which had duplicate controls of the bridge above, but instead of the view screen, they had a clear view of space before them. As they were on a test flight, the main viewer was set up so they had clear views of all around the ship and right now they could see the rear of the ship was very close to four of the massive cranes that lifted parts for the exterior of the hull into position.

"Increasing power to the engines, we are now on our way," Miss. Jackson replied as the ship moved slowly forward and at an angle, then increased speed taking only sixty seconds to attain an orbit of 300 kilometres.

"Thank you," Captain Scott replied and sat back in her seat breathing a heavy sigh of relief. She looked around the bridge seeing everyone was now breathing easier and getting on with their jobs.

A week later Explorer landed back at its spaceyard, coming in a lot slower and completely level than when it left Earth, but now the crew had a chance to get used to the controls and the ship landed perfectly, right on its spot with the weight of the ship taken by its two thousand support struts.

As soon as it was down, people were checking the outside of the ship for meteor strikes but found none and nothing was wrong with the hull, everything; including the force field had worked as expected. As usual, Mary was there with her top disciples to welcome the ship home and thank the crew for all their hard work and faith in the ship, its builders and designers.

Five weeks later, a month before they thought this would happen, just fourteen months to day after Mary arrived on Earth, five ships lifted off into space taking with them one hundred and ninety thousand people to the stars.

First Lady Rocc and her family, with two hundred disciples were aboard the space Ark Enterprize; First Father Langley with his family and two hundred disciples were aboard the Space Ark Dauntulous. These two ships left orbit within ten minutes of each other, passing the second giant Orbital Space Wheel, Hercules, which had been put into orbit only a week before.

Second Father Huntley was aboard the Excalibur with his family and two hundred disciples and on the last smaller Ark, First Mother Meadow with her family and another two hundred disciples was aboard the Valiant. These two ships left orbit ten minutes after Dauntulous, and as they were leaving, the largest Space Ark Explorer, with Third Father Mathews aboard with his family and four hundred disciples, left Earth's orbit and with them was Mike Adams, who was promised a ship by Mary when he interviewed her on the day she arrived. This would be his greatest trip, one of the first humans to visit another galaxy and meet extra terrestrials who would eventually come to Earth and join the galaxies together.

The first of the Trade Delegation ship, the Firefox, lifted off from Australia, where one day soon it would bring the first delegates from another world back to Earth. The Firefox was seeing the Space Arks off on their long journeys into space and then it would travel to Mars and drop off another eight hundred homes and much needed equipment. It would also take to Mars, another three hundred Trade people to help get the new Parliamentary Complex and Trade Delegation Complex completed.

A week later, the next ten Space Arks all lifted off within ten minutes of each other and once in orbit, waited until they were all ready and then each ship left in a different direction, taking Gods' disciples to new worlds, a long way from Earth, ready to spread the word of the True God, the female God that nobody knew existed until just a short fifteen months ago. Now the whole universe was going to be woken to the new age, the age of miracles, the age of women and the new age of God.

Two months later as Mary watched more of her ships go forth and travel to new worlds; she knew her disciples would do well. She had taught them well, put them through the rigours of tests and knew they would come through for her and her parents, who now ruled together over this universe and the Zones within it. As the Explorer went on its way, Mary sent a final message to Mike, thanking him for all his hard work and dedication to her cause.

"Good luck Mike, I'll see you again very soon when you reach the Spiral Galaxy of Andromeda. You will be able to introduce me to the new people, where once again, we'll make our miracles happen and change the way the people out there think about God."

When Mike picked up his message, he smiled and looked at the large screen in his state room. They had stopped for a short time at three planetary systems and seen some wondrous sites and he had met some wonderful people. These new Trade Delegates were now on their way back to Earth aboard the schooner Firefox, to be the first delegates on Earth to start the Trade agreements which would be the start of the new world and a new life for the people of Earth with Mary, the daughter of God, at their lead.

CHAPTER 22

As the thirtieth space Ark Courageous, left its spaceport in India, a new design, now four kilometres long, two and a half kilometres wide and two and three quarter kilometres high, attained orbit, its crew of thirty thousand people commenced their own search of the stars. They were travelling very far from Earth, heading to the centre of the galaxy where it was hoped they would meet up with the Sagittarians, who were already travelling far into space and had sent a message of greetings and felicitations to Earth to request a meeting with the view to start trade negotiations.

Everyone was very happy at this news and with the trade delegation centre having been made ten times larger than it first was, there was now plenty of room for the next visitors to come to Earth.

When Luciferious was ready to return home he sent a message to his friend Rear Admiral Marcus to come and get him. He also had a large number of people to bring with him who would not be able to leave the way Godisious and Mark had. They had arrived in the early years when the experiment was in its infancy, using a different method and when Luciferious had entered the zone, he had come by ship with the rest of his team, all eleven hundred of them.

Rear Admiral Marcus's ship, the Viscount, had been seen around the Earth a number of times and been seen inside Earth's atmosphere over a five thousand times through the years. This project had been going a very long time and it had been very hard for the families of the people working on it, including Lucifer's two sons, Mark and Jesus and his daughter Mary, who was still on Earth. When she left Utopia Prime to come to Earth, like everyone else, she had forgot her other life and what she had left behind, but Luciferious and his special engineers had not. This was his project and his wife had endured hardship for the past five years, despite the fact that they talked once a month, it was long time. The other thing he had done was to swap names with his wife, it had been the only way around their situation but it had worked perfectly.

With more ships returning to Earth from the outer regions of space, Admiral Marcus used the Viscount's Holo Projectors to make his ship appear like one of the incoming ships from the Alpha Centauri Trade Delegation, which was now going to and from Earth numerous times a month.

The ship came down and landed in the cold Antarctic, where there was a small scientific outpost with a church not far away, something the people there were not expecting to see. The scientists were working away from their base and it was left to Disciple Adrian, who was in charge of the outpost and the church, to investigate why this had landed here. He sent a message to Mary as he had never seen the ship before and he wondered if it was damaged and the people were also different to what he was expecting.

Within ten minutes of his warning to Mary, she arrived at the church and spoke calmly to Disciple Adrian before going to the ship. She stood outside what she thought was the main doors, then noticed something else, armour and weapons, none of the ships from Alpha Centauri carried such armourment, and right now she knew this ship was not from Alpha Centauri.

She tried to see inside the ship but was met with a much stronger mind than hers, and she wondered who it could be. She called her Mother and then another thought entered her head, it was her Father.

"Mary, he said standing behind her," he laid a hand on her shoulder and she turned instantly to face the man she known all her life.

"Father, this ship, hold on, what are you doing here? I thought you were going home to be with Mother?" she asked sounding concerned.

"Yes I am, but things have changed and for me to return home, I can no longer travel like I could before and so I had to arrange for my own special transport, as too for my fellow special angels," he replied.

Behind him people were climbing out from beneath the ice field. A door was opening from the side of the ship which was now showing its true size and armourment.

"This is a warship, what is it doing here?" Mary asked showing concern in her voice.

Lucifer looked up to the giant ship seeing the actual size and her name appearing before them.

"Quickly, get aboard," Lucifer said to his people.

"Father who are these people; I thought all the souls were being returned to Heaven?" Mary asked.

"They have been, these are the last of my people who needed to come with me, and they wanted to see what your mother had created on the trip home."

"But why don't you teleport home?"

He looked at her with so much love and care in his eyes, it is not possible from the surface; all will be explained to you when you return home. How long do you envisage remaining here?"

"As long as it takes," she replied.

"One thing to remember, do not stay too long, you can fall in love with this world, it has a bond over you, but make sure you keep the women in command of the ships, when the men were in charge, all they wanted to do was take their wars into space, for the moment, you must make as many friends as you can."

"If I may ask father, how long have you been here?"

"Here, as on Earth, around 4500 years, since the project was started. How long, seriously are you thinking of staying here?" he asked again.

"At least a hundred years, I want to ensure my work is moving in the right direction and everything will go according to plan."

"Are you remembering anything now?"

"The ship, it seems familiar to me," she replied as she looked again at the ship as more of the hologram disappeared. More people were climbing from deep underground and quickly getting aboard the ship. They were running because it had been hot underground and out here, in the snow and ice, it was freezing cold and they all had lots of equipment to carry.

"As time goes on, you will recall this ship and . . ."

"It's the Viscount, I arrived here on Earth in this ship but it had a different shape then, or at least appeared to, it had the shape of a . . ."

Her father looked at her and smiled. "A church, Radstock Church to be precise."

"Yes, that's right. It's all coming back to me now, I'm a Utopian and this planet, this cosmos is inside a," she paused for a very long moment, watching the people get aboard the ship. They were normal people like her, tall with blond hair and electric blue eyes. She looked at one of the men and the woman behind him.

"Yasmin," she called to the woman. "Yasmin, is that you?" she asked.

The woman looked at her and smiled, her eyes turned to Luciferious and then back to Mary. "Yes, it is I," Yasmin replied.

"But you were on Utopia only a year ago then you left on a mission for . . ." Everything was coming back to her like a train crash.

"This planet, this cosmos is inside a Hadron Collider that you built with help from Messengerious, Gabrielious, Opportunious and mother, Luciferious, but you and mother have swapped names, because your name is Godisious. This planet, the stars, outside it's all so small and in here, so big," Mary said.

Yasmin walked over to Mary and touched her hand. "It's nice to see you again Marisious," she said using Mary's real name. "For us the mission is now over, we have all had to work very hard to get this planet moving in the right direction and for a time, we thought the people would never achieve space flight. But now with you helping everyone, it was indeed the right choice. After all, your father has given five years of his time on this project and we have given between one five years. If you stay for a hundred years, it's only 33 days, like taking a trip to the Browlean star system.

"The one thing I can promise you, you'll enjoy every moment you're here, especially with the powers you've been given, but use them for good and don't get too carried away or your mother will tell you off. At least now you know that your father will be back home with your mother and we need him there. Time is quickly passing; it will not be long before the Cycloves arrive in our galaxy and we must know by then how to defend ourselves or what to say to them. It will only be about a hundred years before the Cycloves we added in the 13th universe here, discover their way out and into the next universe.

"They are already making many allies and killing thousands of other species as they make their way through their galaxy. You must get your ships out in this galaxy to make as many friends as possible and spread the word of God, which is what will join you together and help defend yourselves in the future. Do not allow the weapons of mass destruction to get into the hands of the men on this planet for many years. It will take you at least sixty years to make enough friends and bring them back here and spread the human race out into space. Oh there is something you can do for me, as you will be staying here for about a hundred years."

"What is it?" Mary asked.

"In about a hundred years time, a man called Rydian Chord will get together with a woman called Ursula Hammond, they have to marry and move out into space on one of your World Ships. Make sure they go to the planet Amadalem; which is in the Plato Mibus star system, a blue star, three hundred and ninety seven light years from Earth. They will have a son called Kylan and he is to marry a woman called Madelyn Rhysco. Can you please make sure that somehow, these two people get together and remain together?"

"I will do my very best, but why are these people so important?" Mary asked.

"I have been working here on how future people's lives will have an influence on how we get the answers to the questions we so desperately need. The answers as to why your father built this Hadron Collider, for without it, we would be lost and the future of all the worlds inside this Hadron Collider depend on these four people getting together and the children they will have," Yasmin explained.

"I will carry out your wishes, do not worry, I have a very good memory and I will not forget your words, or this time we have had together here."

Yasmin leaned forward and kissed Mary on the cheek. "Goodbye my love, take care."

"I will aunty, bye for now," Mary replied, then realised what she had called Yasmin and realised something else had just entered her memory.

"I'm sorry my love, you seeing this ship should not have happened, it was not meant to be. When we built the Hadron Collider we missed out a very important item which has meant we have had to resolve to use a ship to get us in and out and to Earth especially. Had I thought the problems right through, I would have realised we needed a door or portal to allow us in and out without having to go through the dream process. As we don't have the portal, we have to come in and out by ship."

"What is the dream process?"

"The way we send our soul to the people here. Normally you are asleep on the other side, dreaming, while your soul, for want of a better word, moves the body around the planet. But when we need to come here as a whole person, we have to come in by ship and leave by ship. Is it all coming back to you now?"

She looked at her father again and gave the situation further thought, and then she recalled everything, like waking up from along sleep.

"I remember now, I remember everything, Utopia Prime is our home planet, the Cycloves are a very violent species, my job here, leaving home and Port Troy. I remember everything, will this ruin my job here?" she asked her father.

"Hopefully it will make it easier for you; at least you know who your mother and father are, and your two brothers. Use your gifts well, get the people of Earth out to the stars, get them to meet others and make them strong. Within the next four years, have them make more another four pairs of terraforming machines, and then you will be able to move them around the planets and start moving the people from Earth out to the stars. This is will your main job now, get the people out to the stars, let them explore the galaxy, and keep them thinking of God. You can tell the world that Lucifer has gone back home and left the planet to your mother. Tell your people to have faith in God and move out to the stars with God's pleasure. That way, they will stay in peace and this will give the people of Earth the chance to explore and find new worlds on which to live and expand the human colonies.

"There will plenty of time in the future for warfare and war will come, but not now, not for the next ninety years anyway. This will give you the time you so desperately need to spread your wings. I will have a word with your mother and get her to gift you a further hundred large World Ships and the very latest designs in engines. Ah here is he comes, do you remember this man?"

A tall man walked down the ramp from the ship to meet Godisious and Mary. He held out his hand to take Mary's and shook it with great enthusiasm.

"You are," she paused a moment looking at the man standing before her wearing a light blue uniform of the Utopian space fleet. "Rear Admiral Marcus?" she suggested.

"That's right Mary; I brought you here to Earth when you first set foot upon the planet. I can tell you, it was a hard job getting the ship to resemble the church, but the crew loved all the pomp and circumstance that went with getting you from the ship to the planet. Luckily the sun, being so bright managed to hide our ship as it descended through the cloudless sky, which was another good idea of your father's as we could merge our ship very easily with the blue sky and not be seen from the people on Earth. I must say, we all enjoyed your day, and we have it on record and sent everything to our people back home. I will say it was a brilliant idea of your father's to send in the tanks, we were all in stitches wondering who would be the one to stop them."

"Actually it was Verioncaious who came up with that little plan, just to keep the police happy and show everyone on Earth that God was protecting them. I think it was an excellent idea, what did you think about it."

"I thought mother handled it very well, the people certainly all believed in her more when the incident happened and were grateful she was there for them. You were losing followers every minute of the day as it progressed," Mary said, sounding sorry for her father.

"That was the whole idea, this had to be the end of Lucifer and the beginning for God and for the people of Earth to go out to the stars and explore new worlds and expand the human race. If not, then when the Cycloves get here, you will be wiped out in a few days. We need to see how the humans sort them out, and in doing so, it will tell us how we will either be beaten by the Cycloves, or get them to stop fighting all the time killing billions of lives throughout the real universe."

"I understand now, it is very important that the human race expand and move out into space, I can remember the whole project now, and the reasons why you had to come here and get this planet started and move the people of Earth forward, for these are the species most like us, and the Cycloves, although much higher up in the Hadron Collider, will in the very near future, break out of their universe and come searching for us.

"If we don't want to die, then we have to get this experiment to work and that is my job, to get the people out into space and meet other people who will in time, help us create new weapons of war against the Cycloves. I'm glad I met you now, if I hadn't, then I may have forgotten why I am here."

"Your mother would have told you in your dreams or using telepathy, which was her doing not mine. The telepathy aspect was your mother's idea, for when we started this experiment, it was not thought of, we didn't think about contacting anyone inside the Hadron Collider as we knew they would be gone for only two weeks at the most. When I realised someone would need to be here all the time to get the people moving, the Hadron Collider had been built and there no way we could get our thought patterns to reach here. Now, you have a transmitter inside you which sends your thoughts to Port Troy and they are then sent outside of the Hadron Collider to another transmitter which sends your normal telepathic thoughts straight to your mother, wherever she is in our universe."

"Yes, I remember now, can I speak to you Rear Admiral Marcus using telepathy?"

"Yes, now that you know me, when I am inside the Hadron Collider, you can in future contact me and I can reply to you. If you need a ship or need to get somewhere fast, let me know, it will not take us long to get to you and we could use this base as meeting place. It's cold, out of the way and apart from the small base and church, there are very few people here."

"Which reminds me, I must speak to the people here and Disciple Adrian to let them know you are not aliens come to kill us."

"What will you tell them?"

"Being honest I have no idea, but I'm sure something will come to me when you go," she paused a long moment thinking of the old bible and some of its stories, of the books written about God in days gone by, one in particular caught her long term memory, 'Chariots of the Gods'

"Actually, when you lift off, I know you don't use fiery engines, but can you do that, using the holo projectors, make the ship look like a Chariot of the Gods?" She formed a picture in her head of what she meant and sent it to her father and Rear Admiral Marcus.

The two men looked at each other and smiled. "Yes, I'm sure we could do that, I'll get my crew working on the idea immediately," Rear Admiral Marcus replied grinning like a Cheshire cat.

As Godisious turned and looked behind him, the last of the people were now leaving the entrance from below, carrying the last of the equipment which would need to go with them. Six days previous, during the blackness of night, at the centre of the South Pole, the Viscount landed and took most of the heavy equipment from a large opening which led to the depths of Hades, the underground fortress where the people from Utopia Prime worked together getting the people on Earth to move in the direction they would need to go to make rockets, weapons and generally get themselves into space.

Two thousand other scientists and engineers left that same night, taking all their clothes, and material objects they had collected from Earth with them, especially many of the trophies and pieces of art from Picasso and Rembrandt and other great art masters. The ship was down for four hours as everyone worked frantically getting everything from their homes which they had occupied for the past 4500 years.

When the last of the people were aboard, the Viscount took off as silently as it had landed and left Earth meeting another ship the Pleiades, four hundred light years away where the team took all their equipment and transferred to the new ship. The Pleiades, then headed on a five day trip out of the Milky Way Galaxy and once in intergalactic space, up the side of the Hadron Collider to Port Troy, which was situated half way down the Hadron Collider which was in reality, ten thousand kilometres long and a thousand kilometres in diameter, a huge cylinder, floating in space over a moon and protected by a heavy force field from meteors and ships that got too close.

Once through Port Troy, the ship expanded to its normal size, four kilometres long, real time kilometres, not the reduced size it was inside the Hadron Collider. It then took the crew back to their home worlds to meet their loved ones who they had been separated from for just over four years, some of them for five years, like Godisious had been.

Now the Viscount was back to take the last of the scientists from Earth back home, there would be no transfer for these people, they were leaving Earth and heading straight for Port Troy and then home to Utopia Prime.

Godisious put his arms around his daughter and gave her a long close hug, then whispered into her ear. "I love you, take care," he said and then they parted.

Rear Admiral Marcus also gave her a hug and kissed her cheek; then he departed for his ship.

"Goodbye for now my dear, I'll see you in a few weeks of our time, don't forget, a week home is thirty years roughly here, so you won't be gone that long. Do your job well, I will still be able to see what you're doing, and oh, I just thought, before I go, take this pill."

He took from his jacket pocket a small plastic case with a yellow, oval pill inside and handed it to her.

"What is it for?" she asked.

Swallow the pill whole, it will make its way into your blood supply and end up just outside your stomach, this will allow you to speak to me on the outside of the Hadron Collider when I am home on Utopia Prime, it's similar to what you have injected into your arm so you can speak with your mother, this pill will allow me to speak with you."

"If we hadn't met, how would you have got this pill to me?"

"I would have given it to Disciple Adrian, and asked him to give it to you before I left, talking of which, he is coming towards us. So I had better say goodbye and take my leave."

"Okay father, take care of yourself, I will ensure I do as you ask and also what Yasmin asked of me, it will be done. Don't forget the Chariot," she said smiling.

"I will, oh here he is."

"Mary, are you alright, I have seen you speaking with these people and their ship appears to be a ship of war, have they come to attack us?" Disciple Adrian asked sounding petrified. He was used to seeing large ships, so the size of the ship didn't worry him; it was the weapons of destruction that it carried that concerned him.

"Have no fear Disciple Adrian, all is well. I am very pleased that you summoned me here. This person who stands before is Lucifer, denounced of his powers and now just a man like you. He is no longer able to command people to do his will or bring the dead departed souls to his underworld of Hell. I am very pleased to inform you that my Mother Godisious has taken away his powers and he is now left with no other way to leave our planet except by a standard spaceship from a race which lives fourteen universes away.

"God has sent this warship here to ensure that Lucifer is not harmed on the way to Heaven which he will have to travel the long way round, through many Black Holes and through Time itself. The other people in white who you have seen taking their meagre belongings with them are Lucifer's closest disciples who have been commanded by my Mother to return to Heaven and face the consequences for their actions on Earth.

"This is the type of ship that Gods have always travelled in while moving between the planets if their powers had been so low or stripped from them. Since the people of Earth have changed and now pray to my Mother, the True God, Lucifer has lost all of his powers, he is no more than a mortal, but he still commands my respect, for he is after all my father, who just went wrong and made a silly mistake," Mary explained, hoping she had not offended her father.

She looked down at the plastic case in her hand, opened it and swallowed the pill so her father could see she had taken it. Disciple Adrian looked at the man standing before him and the last of the people now walking as if disheartened up the ramp to the ship. As the last of the people came out of the underground shelter, the door was closed and sealed. The surface looked like it had before, covered in snow and ice with nothing beneath.

"Lucifer, the father of Mary, I know you have done wrong, and now you are just like me, but you are and have been a huge part of our lives, could I, if possible, at least shake the hand of the father of Mary, the daughter of the true God?" Disciple Adrian asked.

Lucifer looked at him shocked for a moment, then held out his hand, reading his thoughts to ensure he was sincere in what he had said and meant no harm to him. Disciple Adrian took his hand and shook it.

"Thank you for the honour Sir," Disciple Adrian said and took a step back to be by Mary's side. "I am pleased you are leaving us and we are now in the hands of the True God, with her in charge, I am sure we will travel far into space and make many new friends."

"I am sure you will; now Disciple Adrian, my captain calls me to join him aboard his ship. I ask you to take care of my daughter and follow her for all time and do as she asks. I was wrong to do what I did, to make man move away from the real God, but you are now on the right course. I wish all humanity well and I hope you will all travel far and wide and see the many wonders my wife has made for you."

When they looked up to the top of the ramp, a man wearing a captain's uniform was looking down at them.

"I will hold you up no longer Lucifer, goodbye I'm sure now with Mary to guide us and following the real God, we'll go far."

"I'm sure you will."

Disciple Adrian turned and walked away from Mary back to his church. Godisious kissed his daughter one more time then turned and followed the last woman up the ramp. At the top of the ramp, he turned and waved goodbye to Mary and the ramp withdrew into the ship and the doors sealed. As Mary walked back towards the church to join Disciple Adrian, they both watched as the ship lifted off with fiery flames coming from beneath the ship and as the ship tilted and started to climb, more flames came from the stern.

Inside Mary smiled, it did indeed look just like a Chariot of the Gods, and Disciple Adrian watched with her as the Viscount accelerated up through the clouds and away from the Earth. As soon as it was out of Mary's sight, the holo projectors were turned off and the full cloaking device turned on to get them through the congested lower atmosphere where shuttles went up and down and to and out to the Moon and Mars, and now, further out to Jupiter and Saturn.

Already minors were working in the Asteroid Belt, collecting new minerals and ores to make their new ships from and on one of the asteroids, a man called Ghazawan Albu Yaser and his wife Umria, were drilling deep into one of the biggest asteroids they could find. It was out of the normal asteroid belt, falling some way behind the inner belt, but it was big and their small ship had all the latest lasers and drilling equipment to get the material out of the asteroid.

There, they discovered five hundred large cases of highly advanced weapons; they had no idea how long they had been there, but it was now theirs and they had not reported it. Four other members of their family knew of the advanced weapons, which they had all used and made the hole into the asteroid much larger so they could get their ship inside it; well hidden and out of sight of prying eyes. As they wondered what to do with the weapons, the Viscount passed by the asteroid and detected their ship.

Parking close to the asteroid, the crew watched, seeing the mauve laser beams leaving the inside of the asteroid and travelling harmlessly into space. When Godisious was called to the bridge, they discussed what to do about the people.

"These weapons were not to be found for another sixty years," Godisious explained.

"So what do we do with the people?"

"Let's put them to sleep for the next sixty years, if they are discovered, we'll make it appear they hit a gas pocket, and when they wake up, they will know the people of Earth needs the weapons and take them there. They will assume another race has attacked them and they are to take the weapons back home as fast as they can. They'll end up being heroes, rather than terrorists; what do you think?" Godisious asked Rear Admiral Marcus.

"An excellent idea, Commander Stewart, can you arrange that for Godisious?"

"Of course Sir, I'll get onto it immediately. We should be able to get underway within the hour," he replied and taking one man from the bridge with him met the rest of his team in the main hanger.

An hour later, with Ghazawan Albu Yaser and his family sound asleep inside their craft and the ship's chronometer set to activate the equipment to bring them around in sixty five years, they sealed the main entrance to the asteroid hiding their ship. It would be very simple for the people to open the entrance; one pulse from the laser, which Commander Stewart had left in their ship, would set them free. They would read the note which was left for them and race back to Earth taking the weapons with them and hand them over to Mary.

Five days later the Viscount arrived at Port Troy and when it went through the airlocks, it emerged into real space outside the Hadron Collider. As soon as the ship was at a safe distance, it started to increase in size as they moved closer to the planet Miran, which was the next planet to Utopia Prime. Miran's largest Moon Hiabus was now before them as the Viscount turned so its main front cameras could see what they wanted. As the ship continued to move forwards, away from Port Troy the scientists and Godisious once again could see Hiabus and just above it, the huge Hadron Collider in its true form.

It was hard to believe they had been inside the object for five long years of their time, which was 4,500 years in Earth time. A short time later the Viscount arrived in orbit around Utopia Prime and with clearance given, landed at the main spaceport.

When everyone had left the ship and the experiments and equipment had been unloaded, Godisious walked into his house and met his wife Luciferious. They embraced and kissed each other before talking and catching up on old times and what she had been doing to move them and their people forward with the Cycloves not far from their own galaxy.

"I am sure with Mary in charge they will now move forward and expand as they should be doing. It was hard to get the people motivated, but once they started to move they did, but they are so obstinate we will have to help Mary by sending people to the planet with ideas already in their heads to move the people forward and get them to the next phase of the experiment, designing weapons of war. However, on our way out, we noticed that one family had discovered our first set of advanced weapons and were planning to use them."

"I hope you managed to stop them," Luciferious replied.

"Oh yes, they are fast asleep inside their ship and will be woken in sixty years and hand the weapons over to Mary."

"Yes, and we do not have long, the Cycloves are only three months away from our galaxy," Luciferious said.

"Mary will have the humans fully integrated during the next ninety years and have enough friends to help them build a big enough war fleet to see how they cope with Cycloves in their galaxy. Even if it takes them a hundred years to beat the Cycloves, we will still have time to prepare ourselves for the battle which will eventually come to us," Godisious sighed.

"Let us hope so, let us truly hope so my love. I have heard from Mary, she has told me about your talk just before you left and she said to tell you Disciple Adrian has not stopped talking about the Chariot of the Gods and she will do her best to ensure everything she has to do goes as planned. She has total recall now doesn't she?" She asked her husband.

"Yes, but that is not a bad thing, now she will know what is really happening and how important her job is. She promised me she would not let us down and I believe her."

"So do I, but she is so far away."

"She is only as far away as I was for the last five years and you can see where she is every day of the week; come, let us look together."

They walked hand in hand to their rear garden joining Jesus and Mark, who looked up when they heard their father. After saying hello again and hugging each other they talked for a short time laughing and joking about what had happened on Earth. It was now getting dark as they looked up into the night sky, they could see in the distance, the huge Hadron Collider, its warning lights blinking on and off, telling everyone where it was, high in orbit above the Moon Hiabus.

"You see, our daughter is just there, inside the Hadron Collider, a few thousand kilometres away from us, and for her, hundreds of thousands of light years," Godisious said calmly. He looked at his watch and realised it had been almost a day since he had left the Hadron Collider.

"Do you realise my love, in the short time it has taken me to return here from the Hadron Collider, sort out my equipment and get here, a mere short twenty hours, three years have passed on Earth for Mary. By tomorrow evening, six years will have passed and the people of Earth will be ready to land on their first new planet and start the expansion of the human race into space" Godisious sighed.

"It's surprising how time flies in different parts of the universe, you are quite right, I'm sure the people of Earth will get to discover the way to help us fight the Cycloves. We just have to put our faith in our daughter," Luciferious replied and looked deep into her husband's eyes and kissed him, something they had not done physically in five long years, but she knew they would soon make up for it.

"Yes, and I'm sure she will do just fine," Godisious replied when they parted from their long kiss.

He put his arm around his wife's shoulder, hugging her close as they looked out towards the Hadron Collider; in a permanent orbit around the Moon Hiabus which orbited the planet Miran; speaking together to their daughter deep within its multiverse; four people looked in, while one looked out; speaking of good times that would soon come to be.

---THE END ---

About the Author

Terence J. Henley, Terry to his friends, is disabled, formerly an electrical engineer and educated up to a degree in physics and maths. He had to retire in 1993 after he developed an Epidurmoid Cyst inside his spinal column and was given three months to live.

He has also designed three spaceships and the first one he sent to NASA, three years later a very close replica of it appeared on the cover of Spaceflight (a BIS publication) which turned out to be the Space Shuttle. He also designed a theoretical Light Speed engine with engineers from Rolls Royce engine division back in the 70's.

His wife is continually at his side helping him through the long and often very painful days and nights. Terry is an amateur astronomer and a member of the BIS (British Interplanetary Society), and he follows the space programme with keen interest. He hopes he will see when the first people set foot on Mars and start a new life there. His other interests are films, Sci-Fi and Thrillers photography, model building and reading, mainly SF and Horror. His favourite author is Steven King.

Terry has written a Sci-fi trilogy and a teen romance and a large number of novelettes. He now also edits the on-line magazine Odyssey for the BIS (British Interplanetary Society) and helped to edit a new anthology by BIS authors entitled Visionary.